Dust and Smoke

AF518320

Dust and Smoke

Air Pollution and Colonial Urbanism, India, c.1860–c.1940

Awadhendra Sharan

Orient BlackSwan

DUST AND SMOKE: AIR POLLUTION AND COLONIAL URBANISM, INDIA, c.1860–c.1940

ORIENT BLACKSWAN PRIVATE LIMITED

Registered Office
3-6-752 Himayatnagar, Hyderabad 500 029, Telangana, India
e-mail: centraloffice@orientblackswan.com

Other Offices
Bengaluru, Bhopal, Chennai, Guwahati, Hyderabad, Jaipur, Kolkata, Lucknow, Mumbai, New Delhi, Noida, Patna, Visakhapatnam

© Orient Blackswan Private Limited 2020
First published by Orient Blackswan Private Limited 2020

ISBN 978-93-90122-86-8

Typeset in
Adobe Jenson Pro 11/13
by Shine Graphics, Delhi 110 094

032465

Printed in India at
Akash Press, New Delhi

Published by
Orient Blackswan Private Limited
3-6-752 Himayatnagar, Hyderabad 500 029, Telangana, India
e-mail: info@orientblackswan.com

For

Sheyphali

Contents

Abbreviations

ABP	*Amrita Bazar Patrika*
BB&CI	Bombay, Baroda and Central India
BC	*The Bombay Chronicle*
BDD	Bombay Development Department
BG	*The Bombay Gazette*
BGG	*Bombay Government Gazette*
BIT	Bombay Improvement Trust
BMC	Bombay Municipal Corporation
BMOA	Bombay Millowners' Association
BOMLR	Bombay Law Reporter
CESC	Calcutta Electric Supply Corporation
CSDS	Centre for the Study of Developing Societies
CSSS	Centre for Studies in Social Sciences
DSA	Delhi State Archives
GIP	Great Indian Peninsula
GoI	Government of India
IDRC	International Development Research Centre
IESHR	*Indian Economic and Social History Review*
IFLC	Indian Factory Labour Commission
ILR	Indian Law Reports
IOR	India Office Records
IPC	Indian Penal Code
Judl Progs	Judicial Programmes
MAS	*Modern Asian Studies*
MSA	Maharashtra State Archives
Mun. Progs	Municipal Programmes
NAI	National Archives of India
OGC	Oriental Gas Company

OUP	Oxford University Press
RCC	Reinforced cement concrete
SNC	Smoke Nuisances Commission
TFLC	Textile Factories Labour Committee
TISCO	Tata Iron and Steel Company Limited
ToI	*The Times of India*
WBSA	West Bengal State Archives

Preface

This is, sadly, an opportune moment to write about air pollution. *The State of Global Air* (2019)[1] reports that this is now the world's leading environmental risk factor, surpassing unsafe water or lack of sanitation that had been the focus of earlier public health measures. It is estimated to contribute to about 5 million deaths globally—nearly one in 10, with India and China at the top of the countries affected, reporting nearly 1.2 million premature deaths each. Even more worryingly, the numbers continue to show an upward trend. Deaths attributable to PM 2.5 have increased by 68 per cent since 1990, though more slowly after 2010. On the other hand, while ozone attributable deaths grew by only 4 per cent between 1990 and 2010, they have since jumped to 15 per cent. And India continues to report the highest mortality on account of indoor air pollution, with 60 per cent of the population continuing to use solid fuels to cook. Indeed, the unthinkable seems to have happened with breath seemingly turned away from life. Warnings regarding the avoidance of outdoor air are routine, from medical experts and pollution regulators alike. Not that indoor air fares much better, with an entirely new commerce in masks and air purifying machines emerging in India, as it has in China in the recent past. Quite clearly, public health concerns and the state of the urban environment have got deeply entwined, analogous possibly to the symbiosis between livelihood issues, subsistence economy and environmentalism of the poor in rural areas of the global South.[2] And yet, we know so little about the quality of our air, the practices that impact on it, the manner in which we may or may not regulate such practices and, eventually, how this air affects spaces and bodies, across class and gender. This despite the voluminous climate talk that surrounds us and despite the increasing realisation that 'human-beings-in-the-world' are always and without exception present as a modification of 'being-in-the-air'.[3] To be certain, there are excellent accounts of the contemporary politics of air, especially the air of Delhi where much of the recent concern around air in India seems to have been concentrated.[4] The role of the courts, of pollution abatement authorities, of civic associations, political parties and trade unions, scientific establishments, etc., has merited detailed attention, much to our profit. Yet, there is amnesia about how urban air has mattered over the longer history of modernity in India and what remains of this legacy in contemporary times. It is this history of atmospheric modernity that I narrate in the chapters ahead.[5]

This book began in the knowledge of a brief exchange among officials regarding the rather mundane issue of land acquisition at the time of the making of New Delhi. It would be a good idea, colonial officials had suggested, to acquire land east of the river Yamuna too, in order to better control the nature of economic activities that could be carried out in the city's proximity. If only Calcutta had included Howrah within its municipal boundaries, they reasoned, a haphazard industrial landscape would not have developed and consequently no nuisance of smoke either that the city had to relentlessly endure. And so I began to enquire into the history of Calcutta, with a vague idea of writing about the traditions of land-use planning in colonial India and how these may have impacted on the issue of air pollution. This was hard to find though, especially as I moved into the archive of the nineteenth century. Instead what came increasingly to the fore were considered opinions regarding the *futility* of land use planning for mitigating the smoke nuisance. What seemed to have mattered more were imaginations of pollution as economic waste, around which there had developed a series of important interventions, not only in Calcutta but also in Bombay, in 1905 and 1912, respectively, and to an extent in Ahmedabad and Karachi.[6] And so I turned to histories of waste and pollution instead, only to encounter municipal waste *ad nauseam* in one file after another to the relative exclusion of waste or pollution from industrial sources—my real interest.[7] The only available resources seemed to be the annual reports of the Smoke Nuisances Commissions set up in Calcutta and Bombay and the factory reports detailing the activities of inspectors. Gradually, the archive 'thickened' to include newspaper accounts of the perils of smoke, annual reports of the association of industrialists, and visual materials in the form of advertisements for cleaner stoves and the promise of cleaner futures inherent in the use of gas and electricity. But there were still no dramatic events to mirror the 1952 Great Smog of London and no major reversals in the state of affairs in either city, for the better or worse, through which to narrate this story.[8] But perhaps that was only apt, for if there is one thing that has characterised air pollution for a long time, it is the lack of drama—it is what is called a 'slow killer'.[9] Its history, as smoke and/or as dust, I have come to believe, is therefore best narrated as a story of incremental change—of shifts in perceptions, changes in priorities, innovations in strategy and setting of new goals—through which we may appreciate the different motivations that have shaped urban environmental concerns in modern times, ranging from aesthetic preferences to visions of economic progress and the making of hygienic cities.

Equally, I suggest, it is only through such probings that we may distinguish between earlier interventions to manage atmospheric contaminations, anchored in the quest for greater efficiency of production, and contemporary environmentalism focused on sustainability, resilience and environmental justice. There is, at an obvious level, primacy of the visual in this 'air talk', far more in the nineteenth century than in the contemporary times, but prevalent even today when pollution concerns subside with the onset of the first sunny day, the gases remaining but their presence no longer available to our senses! On the other hand, there is now a far greater concern with invisible particulate matter than was the case before, perhaps with a better appreciation of the correlation between 'vitiated air' and bad health and certainly a much better appreciation of needing to define environmental standards against which we can measure progress or decline. Expertise was privileged then, as now, even as the complaints earlier tended to be around more personalised experiences of grime and soot, while they are about asthma, cancer and other diseases now. Governmental actions and local complaints come together in such an account, situating the coupling of the social and the biophysical in particular space and time.[10] Climate change matters today, as does local pollution, but in ways significantly different from the coupling of sanitation and pollution a century earlier. But perhaps most critically, care for the environment is now linked intimately to questions of democracy and politics in ways radically different from the colonial period when controlling smoke was seen primarily as a partnership between two differently situated European elites, administrators and mill managers.

And yet, there is much that persists too between then and now. Nineteenth and early twentieth century accounts are suggestive of the multiple sources that affect our atmosphere, such that addressing any one more than another is a matter of social choice. At its most obvious, this is a choice regarding the relative prioritisation of indoor and outdoor pollution, and the implication of such choices. At another level, it also suggests how from a range of pollutants, some become matters of greater concern than others, with visibility as a key marker.[11] Historical accounts are also suggestive of the ways in which each 'solution' for environmental improvement may end up producing new sources of concern. Incineration, as a substitute for landfills, we learn, always came accompanied with the risk of nuisance on account of the gases that escaped from the chimneys, a history that we ignore at some peril as we set out to build new incineration plants to generate electricity as a measure of 'clean development'. The use of particular fuels and the likely impacts of fuel substitution, the role of residents and experts, the authority of municipalities, commissions and committees, the interplay of technological cooperation and legal sanctions with respect to industry, the impacts of different modes of transportation and the far greater difficulties of coping with domestic smoke are all issues that continue to be important for us to think through today. The legacy of the twentieth century ethic with respect to industrial pollution, rooted in efficiency, utilitarian conservation and elimination of waste—the same ethic that guided natural resource management during the colonial period—also has much to offer to energy debates today, even as more environmentally oriented discourses and practices have begun to alter the ways in which pollution is addressed. All this to consider anew, how can we best anchor our concerns regarding air—as an aesthetic choice for the city beautiful, as a measure of efficiency or as health concern—in order to achieve the greatest measure of justice?

In this book, I look at the varied sources of energy, the persistence of old trades, the organisation of new industrial production, metropolitan and local capital, migrant labour and urban development projects that came together in Calcutta and Bombay to produce new sites of work, habitats and commodities, on the one hand, and smoke and dust, on the other.[12] There was much to be cherished about such developments, the visibility of industrial smoke standing in for progress: 'At times our critics are tempted to be very impatient when they ask us to show where are the factories, the dust, the smoke and the bustle of work as a result of all that we have been talking for the last 10 to 15 years.'[13] On the other, the fact that smoke and dust could easily cross over from one neighbourhood to another, being present not only in and around factories but also inside homes and on the streets, affecting both the rich and the poor alike, is also what made it problematic. There was much that would be common to both Calcutta and Bombay in this regard, both in terms of understanding smoke as a problem and the strategies for the abatement of the same, in differing attitudes towards European mill managers and Indian stokers, in the faith in new sources of power, especially electricity, as the ultimate panacea. But the two cities were also dissimilar in important ways, differing in their geographies, meteorological conditions, sources of energy, institutional arrangements, and the priority areas to be addressed. Calcutta suffered from low wind velocities and temperature inversion, especially in winter when the ground being colder than the air above it trapped the pollutants near the surface.[14] Bombay, by contrast, was famous for its sea breezes that dissipated some of the worst effects of smoke. In Calcutta, the low-hanging smoke generated an intense concern not only with factories but also domestic smoke. Bombay, by contrast, made routine noises about domestic smoke but focused far more intensely on the smoke from smaller industrial establishments and workshops. Calcutta had access to coal in its backyard, especially after the opening up of the Raniganj mines and the development of the railways in the mid-nineteenth century.[15] Bombay had to obtain its coal from afar, via sea from Natal or Wales or via railways from Calcutta, and therefore at a higher cost. On the other hand, while Bombay had access to cheaper and cleaner hydroelectric power from the Tata hydroelectric plant set up at Lonavala in 1911, the prospects of water as a source of energy in Calcutta always remained rather bleak.[16] All this had a significant bearing on the issue of smoke, given the intimate links between fuel use and air pollution, and the coincidence of the Age of Coal with the period of colonial rule.[17]

In what follows, I also offer a history of smoke as nuisance in these cities, framed variously as waste and as pollution. Interventions regarding this, I suggest, were related to the aesthetic preferences of the British administrative elite and an emerging Indian middle class, professionalised through the work of engineers as inspectors within Smoke Nuisances Commissions and institutionalised through concerns of economy and efficiency. Improved utilisation of fuel, leading to efficient production and dispersal at a height, was key to the strategies designed to mitigate the adverse impacts of smoke, of which three broad phases may be suggested: an initial phase between the 1860s and 1890s when municipal arrangements were deemed sufficient to tackle industrial smoke, but which failed as the expertise of the inspectors seconded to the city governments could not match that of the European mill managers; a middle phase around the turn of the nineteenth century when new laws were legislated based upon expert advice, leading to the setting up of Smoke Nuisances Commissions in Calcutta and Bombay, which seemed to have initially worked with the twin strategies of persuasion and prosecution to restrict smoke within well-defined time limits; and a third phase post First World War in which the accent was on utilitarian conservation, scientific advice and education, to the almost complete exclusion of prosecution as a measure to tackle the smoke nuisance.[18] Public health concerns configured the issue too, though more intermittently, and with some consistency only by the 1930s when a direct link was proposed between smoke and ill health, by which time attention shifted away from industrial smoke to that produced by commercial establishments and domestic fires. In this movement, from an initial municipal impulse to the functioning of special commissions, we get the first glimmer of an urban environmental imaginary, far from popular, but with significant implications for how air (and its pollution) were to be managed for a century and more, before being eclipsed by new laws in the 1970s. At the heart of this enterprise was a compromise between general economic prosperity, protection of business interests and public good, outlined at the beginning of the twentieth century by Frederick Grover, the anti-smoke expert from Sheffield, whose report on Calcutta would become the basis for smoke regulation in India for over half a century:

> It is evident ... that the only factors operating for the suppression of smoke are likely to be, firstly, a sense of public duty in keeping the atmosphere as clear as possible; and secondly, legislation... wise legislation should not go so far as to hamper the industries of the country by prohibiting the use of its local resources. At the same time the public have a just claim to such protective measures as can secure improved conditions of life without undue interference with commercial prosperity. In advising remedial measures, I feel deeply conscious of these issues and am anxious that a judicious compromise may be arrived at.[19]

The nature of this 'judicious compromise' forms the subject matter of the chapters ahead.

Notes

1. Health Effects Institute, *The State of Global Air: Special Report*, Boston, MA: Health Effects Institute, 2019, pp. 11–13. The data reported is for 2017.
2. On the distinctiveness of the environmentalism of the poor, see Ramchandra Guha and J. Martiner-Alier, *Varieties of Environmentalism: Essays North and South*, London: Earthscan, 1997.

3. The phrase is from Peter Sloterdijk, *Terror from the Air*, trans. Amy Patton and Steve Corcoran, Los Angeles, CA: Semiotext(e), 2009, p. 93.
4. See, for instance, Armin Rosencranz and Michael Jackson, 'The Delhi Pollution Case: The Supreme Court of India and the Limits of Judicial Power', *Columbia Journal of Environmental Law*, 28, 2, 2003, pp. 223–54; Amita Baviskar, 'Between Violence and Desire: Space, Power, and Identity in the Making of Metropolitan Delhi', *International Social Science Journal*, 55, 75, 2003, pp. 89–98; Rene Vernon, 'Remaking Urban Environments: The Political Ecology of Air Pollution in Delhi', *Environment and Planning A*, 38, 2006, pp. 2093–2109.
5. For rich ethnographic accounts around this theme in China and Hong Kong, see Jerry Zee, 'Breathing in the City: Beijing and the Architecture of Air', *Scapegoat: Architecture/Landscape/Political Economy*, 8, 2015, pp. 46–57; Timothy Choy, 'Air's Substantiations', in Kaushik Sunder Rajan, ed., *Lively Capital: Biotechnologies, Ethics and Governance in Global Markets*, Durham: Duke University Press, 2012, pp. 121–52; Timothy Choy, *Ecologies of Comparison: An Ethnography of Endangerment in Hong Kong*, Durham: Duke University Press, 2011.
6. This does not imply that location did not matter in conceptions of smoke as waste. As we shall see, on numerous occasions, attention was drawn to the fact that smoke was especially problematic in dense urban neighbourhoods while being relatively innocuous in more open areas. On the other hand, what it does imply is that as a strategy, technical change that could either reduce smoke (thus preventing waste) or disperse it at a height was preferred over the strategy of relocation of industries for much of the period of this study.
7. One option therefore would have been to write a 'fuller' account of air in India, concerning both municipal and industrial matters. But several discarded drafts later, I finally opted to focus only on dust and smoke, for purpose of clarity. What I offer instead is an introductory chapter that lays out the distinction between sanitation, ventilation and pollution, within homes and factories, and in the city at large.
8. A recent rendition of events surrounding the London smog may be seen in season 1, episode 4 of the television series *The Crown* in which there is much political drama enacted around it in December 1952 before Churchill breathes a sigh of relief once the sky clears up after some days.

9. This is also perhaps why air pollution has also merited less attention than water pollution, both among contemporaries and later scholars, as contamination of water has been more directly linked to outbreaks of epidemics. Things might be changing now with air pollution seen as an equally big health hazard, at least one that finds greater media and policy attention.
10. I take this idea of the coupling of the social and the biophysical from Anne Rademacher, *Building Green: Environmental Architects and the Struggle for Sustainability in Mumbai*, Oakland, California: University of California Press, 2018.
11. In considering smoke for instance, with which I engage in this book, it was dark particulate emissions of fires that was of most concern. On the other hand, invisible gases that got produced on account of coal combustion, for example sulphur dioxide, failed to garner similar attention. Further, as Stradling and Thorsheim point out, both American and British anti-smoke laws relied on shade for definition and enforcement. Only dark smoke constituted a nuisance that required prevention. See David Stradling and Peter Thorsheim, 'The Smoke of Great Cities: British and American Efforts to Control Air Pollution, 1860–1914', *Environmental History*, 4, 1, 1999, pp. 6–31.
12. The choice of the cities is entirely fortuitous as these were the two cities for which special Smoke Nuisances Commissions were set up and about which we find the maximum information in the archives. Howrah, Ahmedabad and Karachi also find mention, though to a lesser degree.
13. Report of the First Industrial Conference held at Poona, 1891, Poona: Arya Bhushan Press, 1891. Some also believed that smoke was desirable as it had a curative value, especially in preventing cholera. See 'Smoke as a Cholera Preventive', *The Times of India* (hereafter *ToI*), 12 Feb. 1878. In Europe too many believed that smoke acted as a fumigant against airborne impurities, objecting therefore only to the odour of putrefaction and fermentation and not to combustion. Alain Corbin, cited in Peter Thorsheim, *Inventing Pollution: Coal, Smoke and Culture in Britain Since 1800*, Athens, USA: Ohio University Press, 2006, pp. 16–17.
14. Dipankar Chakraborti, 'Calcutta's Environment', in Sukanta Chaudhuri, ed., *Calcutta: The Living City*, Vol. II, Delhi: Oxford University Press (hereafter OUP), 1990, pp. 181–86.

15. Raniganj had India's largest coal reserves, with a total area of about 500 sq. miles and reserves of about 18.60 billion tonnes in the mid-nineteenth century. By the mid-1880s, these fields were linked to all major coal markets by the railways, though the relative prices of freight and shipment determined which coal was preferred in different markets, that from Bengal or coal imported from overseas. Indrajit Ray, 'Dynamics of Bengal Coal Mining in the Nineteenth Century: Dissemination of Mineralogical Knowledge and Railway Networking', *Indian Economic and Social History Review* (hereafter *IESHR*), 52, 4, 2015, pp. 463–99.
16. Government of India (hereafter GoI), *Electrical Undertakings in India* (corrected up to 1912), Calcutta: Superintendent Government Printing, 1913, IOR, V/27/740/4.
17. On the coincidence of coal and colonialism, see Edmund Burke III, 'Preface', in Diana K. Davis and Edward Burke III, eds, *Environmental Imaginaries of the Middle East and North Africa*, Athens, USA: Ohio University Press, 2011, pp. *ix–xii.*
18. Hugh Gorman points out that in this focus on production and dispersal, what got missed out was the movement of contaminants through the environment and their possible interactions therein. This is an aspect of controlling air pollution that would see clear departures only in more contemporary times with a greater focus on the chemistry of air pollution and clearly mandated environmental objectives. See Hugh Gorman, *Redefining Efficiency: Pollution Concerns, Regulatory Mechanisms and Technological Change in the US Petroleum Industry*, Akron, OH: University of Akron Press, 2001.
19. Frederick Grover, *Report on the Abatement of Smoke Nuisance in Calcutta*, National Archives of India (hereafter NAI), Home, Public B, No. 95, Nov. 1904.

Acknowledgements

Some books are chance products of other books, as this one is. It germinated as an idea while I was conducting research for another book on Delhi. That is when I first came across an essay by Michael Anderson on the history of air pollution in Calcutta, setting me off on a search for similar studies of other Indian cities. Much to my surprise though, over two decades and more, Anderson's pioneering essay has remained the only one in the Indian context. In many ways this book is an attempt at filling this void.

The research, writing and publication of the book has benefitted much from the excellent support offered by the Centre for the Study of Developing Societies (CSDS), Delhi. This has been my intellectual home over several years now, and I am hugely indebted to the scholarly engagement and intellectual autonomy that the Centre offers. All the chapters in this book were first presented at the annual faculty seminars of the Centre. To all my colleagues who have participated in these seminars and offered constructive criticism, my deepest thanks. Abhay, Aditya, Ananya, Baidik, Hilal, Nishikant, Peter, Prabhat, Prathama, Rajeev, Rakesh, Ravi Sundaram, Ravi Vasudevan, Ravikant, Sanjay, Sanjeer, Shail and Vijaisri have all been important interlocutors for this book and the ideas developed here owe much to them. My friends Alankar, Ashish, Rema, Sumit, Saumya and Prabhat have been ever supportive and to them too a very special thanks. My teachers Dipesh Chakrabarty and Shahid Amin remain my well-wishers and I thank them for their constant words of encouragement. A special word of thanks also to the members of the staff at CSDS who helped me discharge the administrative responsibilities that came my way while I was writing this book.

I have benefitted immensely from presentations made at workshops and conferences organised by colleagues based at other institutions. Many thanks to Aromar Revi, Dipesh Chakrabarty, Henrik Ernstson, Kshitij Sinha, Rohit De, Sanjoy Chakravorty, Vinita Yadav and Yasmin Arif for inviting me to deliver talks and share ideas with colleagues and students at the Indian Institute for Human Settlements, Delhi; University of Chicago; Stanford University; Indira Gandhi Technical University for Women, Delhi; Ambedkar University, Delhi; Institute for the Advanced Study of India, University of Pennsylvania; School of Planning and Architecture, Delhi; and Department of Sociology, Delhi University.

Avinash and Ritwik, librarians at CSDS, and the staff of the British Library, London, National Archives of India, Delhi, Maharashtra State Archives, Mumbai, and West Bengal State Archives, Kolkata, have provided much needed support. My gratitude to them for their generosity with time and help. Over the past few years, digital repositories have made the task of researchers much easier than before. I am especially grateful to those anonymous persons who have made these excellent repositories—DSpace repository at the Gokhale Institute of Politics and Economics, Pune, Granth Sanjeevani of the Asiatic Society, Mumbai, and the excellent resources of archive.org. The digital collection of *Amrita Bazar Patrika* at the Centre for Studies in Social Sciences (CSSS), Kolkata, was also a major help.

During the period that this book was in the making, CSDS was the recipient of a major grant from the Think Tank Initiative of the International Development Research Centre (IDRC), Canada. This funding enabled me to undertake several research trips to Mumbai and Kolkata and to the British Library, London. Anindya Chatterjee, Kaveri, Samar and Seema at the IDRC office in Delhi have provided not only much needed financial assistance but have also offered their friendship, and for this I am grateful. The Indian Council of Social Science Research in Delhi remains a major funder of the Centre and my thanks to them too for their steadfast support.

At Orient BlackSwan, Insiya Poonawala, Nilanjana Majumdar, Roopa Sharma, Samprati Pani and Vidya Rao have been both patient and diligent. They have been a delight to work with and have nurtured the book with care. Comments from anonymous referees helped sharpen many of the arguments of the book. Many thanks to all of them.

As always, the biggest support in writing this book has come from my family. To my eternal regret, my father, Shrivallabh Sharan, passed away before the book could be published. However, his steadfast support, enthusiasm, gentle queries and, above all, his jest for life till the very end, remain abiding memories. His imprint on me extends much beyond this book! My mother Sushila, and my parents-in-law, Sushma and Kavi Dutt Bhargava continue to provide boundless love and care. To my brothers, sisters, in-laws and their families—Arvind, Indu, Chhavi, Aakash, Ayansh, Kshitij, Meenakshi, Ashish, Divya, Amiya, Yash, Arvind, Shreeja, Mrigank, Nikhita, Pratinav, Naveen, Kumkum, Varun, Vaibhav, Raju, Madhu, Manik, Manya, Vibhu, Romonika, Arunima and Ananmay—a heartfelt thanks for all the wonderful times and unending conversations.

My children, Aniruddh and Maitreyi, have subjected me to many questions, offered several design ideas, provided much needed technical support and remain my most critical audience. As they prepare to step out into the world to make their own careers, their questions and opinions are wide-ranging—from the arguments of this (and other) book/s to the fonts and colour schemes that make for good book covers and from political developments globally to the challenges of institutional life in contemporary times. For helping me appreciate the many different ways in which words and images can be enjoyed, and for being exemplary 'argumentative Indians', I am truly grateful.

Sheyphali has been an indulgent partner in this journey all along. In the age of Twitter, one needs incredible faith to live with the making of a book of nearly 1,00,000 words! For her constant support, her patience, and for a joyous life together, I dedicate this book to her.

1 Introduction

Sanitation, Ventilation, Pollution

The air of cities in India, as elsewhere, has always been suffused with matter. To the residents of these cities—the old towns of Banaras or Delhi or the new presidency towns of Calcutta and Bombay—these matter caused the air to have a specific feel about it, either laden with moisture or dust in particular seasons or smelling of incense or dung on a routine basis. This 'feel' of the air, however, did not organise a hierarchy of civilisation in pre-colonial times. It was just as things were, to be careful about or profited from. In a series of letters written by François Bernier, travelling through Mughal India in the latter half of the seventeenth century, we get a sense of these significations of air as they would have been perceived in medieval times. Writing from Delhi on 14 December 1664, Bernier noted the onset of the Mughal king Aurangzeb's journey to Lahore and Kashmir 'in order to benefit his health by change of air', prompted perhaps by his sister Roshanara Begum, who had been 'long anxious to inhale a purer air than that of the *Seraglio*'.[1] A couple of months later, writing from Lahore where the Mughals had set up a temporary camp, Bernier cast his glance away from the preoccupations of the royalty towards the rather quotidian activity of the making of the evening meal by ordinary people:

> But the greatest annoyance is perhaps in the evening ... This is the time when the common people cook their victuals with a fire made of cow and camel dung and green wood. The smoke of so many fires of this kind, when

> there is little wind, is highly offensive, and involves the atmosphere in total darkness.[2]

Subsequent commentators on Bernier's travel diaries put a subtle gloss on his vocabulary, the *Oriental Herald and Colonial Review* (1826), for instance, translating Bernier's comments into the language of 'nuisance': 'those who walked out of their tents in the evenings were subjected to one grievous nuisance—the smoke and stench arising from the innumerable fires of cow and camel dung and green wood, which the common people kindled about that time to cook their victuals.'[3] Nuisance was a term of ancient usage, designating practices that were seen as damaging to the value of an adjacent property, but in this translation of a medieval descriptive account, I suggest, we also begin to detect a new hierarchy of civility and modernity, anchored in stench, smell and smoke, to complement that organised through geographies of waste, distinctiveness of habitats, and customs and manners of the people.[4]

In this chapter, I intimate the making of this new hierarchy in Indian cities, in ways that made them distinct from Euro-American cities.[5] In doing so, I also seek to offer an intervention in the emerging literature on urban environmental history, with its focus on the ways in which nature is urbanised and the implications of such urbanisation for social relations, resource conservation, and the pollution of soil, air and water. For far too long, historians suggest, we have seen nature and cities as antithetical entities, where to preserve the first necessarily meant to limit the latter.[6] In the Indian context, this was further conflated by the distinction between the rural and the urban, with the former usually celebrated as the ethically more desirable habitat and the latter viewed as inherently corrupting and wasteful.[7] What urban environmental studies and urban political ecology propose instead are accounts of cities as socio-natural entities, combining biophysical and social elements, or as ecosystems, with the flow of resources into and out of them mediated by complex economic, political and social dynamics, which in turn have an impact both on proximate neighbourhoods and on distant hinterlands.[8]

It suggests nature as extending into cities, to homes and to sites of work, where experts, planners and residents decide on air quality, clean water, the use of open spaces and the like in the context of specific social and political relations, thus marking cities themselves as environmental spaces.[9] It also suggests nature as being refashioned in cities, changing its qualities and scales on which it can be the perceived and acted upon, both inside homes and within factories and outdoors, on the streets and in neighbourhoods, oscillating between sanitation needs ordinarily understood as the need to deal with filth, garbage and miasma, and nuisance/pollution that was a consequence of inefficient fuel use and thus characteristic of modern industrial life.

The historical (re)shaping of urban environments was also sensitive to its context, of the difference between Europe and its colonies. This required the simultaneous management of populations and environments to achieve 'civilization', 'progress' and 'modernity' in the latter as in the former, while being cautious about the limit of what was actually possible to achieve in the colony.[10] 'No climates exist that are uniformly hot and dry, hot and moist, cold and dry, or cold and moist', wrote James Ranald Martin in his authoritative medical topography of Calcutta. 'Yet certain countries have such a preponderance of one or other of these qualities as to give a very marked character to the physical and moral nature of man.'[11] Half a century later, W. J. Simpson, long-time Health Officer of Calcutta reiterated this sense of difference, bringing together the climatological and the urban:

> Every old town presents examples of the very definite evils which have arisen from streets and houses being permitted to be built without regulation. But if these conditions are common in European towns, they are still more common in tropical countries and to a much greater extent. The narrow streets, the winding alleys, the crowding together of houses, form an insanitary labyrinth, which cannot be efficiently cleansed nor purified by a free circulation of air.[12]

This concern with miasma and stench, I suggest, combined with concerns around dust, temperature and humidity inside homes and within factories, to introduce a new sanitary paradigm in the cities of colonial India centred on ensuring good ventilation. There was, in addition, a differently oriented discourse of nuisance and pollution that was a consequence of coal use and smoke, which addressed the question of air purity through a focus on resource conservation, economic efficiency and public health.[13] The phrase 'vitiated air', in other words, was expressive of a double coupling—filth/ventilation, on the one hand, and smoke/pollution on the other, with the category of waste serving as a hinge. Waste, in the sanitary discourse, concerned its materiality—waste as filth, garbage and excrement—which needed to be *disposed*, though the manner in which this disposal happened could itself be productive of smoke (or indeed of pollution of other kinds).[14] By contrast, smoke as waste signified the absence of efficiency, of the failure to *consume* fuel (mostly coal) in a manner that could extract maximum energy and minimise emissions, thus failing to realise the greatest value.[15] There was also a third aspect to this question, of things that were being wasted on account of smoke, most notably declining real estate values and degeneration of human life in dark, ill-ventilated cities.[16] In brief, while the 'corruption of air' in sanitary terms concerned the presence of rotten vegetation and miasma, as pollution, it was about the concentration and dispersal of unburnt carbon.[17] Wasteful use of fuel was always entirely about the economy, but waste when translated into smoke (and urban pollution) concerned a range of things—from health to the economy and from domestic labour to the aesthetics of built forms. And while the sanitary city necessitated large engineering works to deal with sewage and water supply, solutions to smoke nuisance lay in the use of relevant technology by individual firms.[18] Waste remained a shared category across both sanitation and pollution, but it was variously articulated. In what follows I elaborate upon these framings, sites and materiality to offer an account of air in Calcutta and Bombay that briefly alludes to the familiar histories of sanitation and public health, but, in the main, is oriented towards understanding smoke as waste and as nuisance/pollution.

Sanitation and Ventilation

Homes and Streets

Modern sanitary science, born in mid-nineteenth century England to deal with the adverse outcomes of urban and industrial revolution, was directed most prominently at spaces of miasma—cesspools, dunghills, burial grounds, slaughter houses, marshes, swamps and the habitations of the urban poor, where gases and vapours were to be routinely encountered on account of filth, rubbish and decomposing matter.[19] Sanitation was both a 'natural' and a 'moral' issue, addressing health, crime and habits simultaneously. Epidemics, especially cholera epidemics, were an ever-present threat and, what was more worrying from the perspective of the reformers, did not respect class boundaries, easily moving from the hovels of the poor to the mansions of the rich. Sanitary reformers in England spoke of this routinely, their views on the habits and habitations of the English working class translating in the colony into a perception of the Oriental 'other'.[20] The 'civilising mission' of colonialism would thus be about the recognition of the difference between metropolitan cities and those in the colonies, and improvement of the latter as a mark of 'progress'.[21] Indeed, in Calcutta, this was evident from the very inception of the city settlement by the side of the river Hooghly. A more unhealthy place could not be chosen around the river, Captain Alexander Hamilton famously declared, being near a salt water lake which flooded annually and where the putrefaction of the fishes filled the air 'with thick stinking vapours'.[22] Marshlands and thick vegetation, producing constant and unwholesome exhalations, retained and accumulated in a humid atmosphere, contributed to troop mortality, while dirt and rubbish permitted to lie before the doors of the inhabitants 'in a most slovenly and offensive manner' were a more immediate complaint, bringing the natural and the human together in descriptions of urban insanitation.[23] The crooked lanes permitted neither sun that could 'rarify and elevate the vapours' nor wind that could 'dilute and disperse them', a situation that was ripe for ill health.[24] And so, 'the first object', the Lottery

Committee[25] observed in 1817, 'ought ... to be to penetrate this space thoroughly with wide roads as nearly straight as possible and in the direction of the prevailing winds, so that they might serve as channels for the free transmission of currents.'[26] Research in England had already confirmed the excess concentration of sickness and death in courtyards, dens, blind alleys, cul-de-sacs, winds and closes, suggesting open streets—paved, sewered, swept and open at both ends for the free through passage of air—as the solution to the same.[27] This had to now be implemented in the colony too, where also links between dust and disease were being posited, by Europeans and Indians alike:

> When ... we see the dusty and dirty state of the roads and streets, and inhale those noxious vapours which rise in fourfold proportion near the houses and offices of the European inhabitants, ought we to be surprised at the severe causes of fever, or the sudden attacks of cholera![28]

Streets apart, there was also the question of houses, with much being made of the stench and the filth of the *bustee*s and the bazaars. Descriptions of bustees, with their 'green and slimy stagnant ponds, full of putrid vegetable and animal matter in a state of decomposition' whose bubbling surfaces, under a tropical sun, exhaled 'noxious gases, poisoning the atmosphere', continued to speak of 'imperfect' sanitation right through the nineteenth century.[29] In addition to human waste which found its way into drains and tanks, to undergo 'dangerous decomposition', there were other kinds of filth—house refuse, rotting vegetable matter, remnants of food, stables and horse dung and litter—that were removed 'very imperfectly and rarely, or not at all ... to fester in stagnant drains'. Nothing short of a radical and complete obliteration of such 'foetid pits and troughs' would remove this 'palpable and indisputable danger to the public health', observed the Health Officer.[30] But far from this being so, the plan or construction of the city continued to remain 'radically defective', and in a remarkable reversal towards the closing years of the century, the same bustees that had attracted much adverse comments over the years now seemed almost benign—as unwholesome as they may have been, they still permitted wind to permeate through them. By contrast, the groups of big houses, 'separated from each other only by narrow, tortuous and foul alleys', could only obstruct the proper circulation of air, 'with all their attendant evils, foul and inefficient drainage and pollution of air by sewage, by sewer gas, by unremoved filth from defective scavenging and by the organic products given off from animals and men living in a crowded state'.[31] Quite clearly, while much had been achieved by the provision of clean drinking water, clean air was yet to be secured as the necessary complement of the urban improvement project in Calcutta:

> those parts of the native town which are so badly ventilated are in fact, in the matter of air-supply, in much the same position as the whole town was when dependent for its water supply on stagnant tanks. If the drinking of polluted water is injurious to health, so most assuredly is the breathing of impure air, although the effects of the latter are more insidious and less easy of detection.[32]

Large-scale demolition of slums and the making of wide streets were the solutions proffered, with Lieutenant Governor Mackenzie (1896) mincing no words in saying that the chief problem was 'to clear out the pestilential quarters' and 'drive broad roads through those quarters, and to replace those horrid pigsties ... by respectable, well-sanitized buildings'.[33] English observers wrote enthusiastically of a 'new' Calcutta: 'All the dirty spots are being torn out of its heart and flung aside: narrow streets, congested quarters, slums—all. Whole streets are being laid anew, the dust held fast beneath large slabs of tar, so that the city is cleaner and less dusty than it was before.'[34] Long reports were written by members of the Improvement Trust, outlining among other things, nine separate, comprehensive street schemes for the city. As before though, these plans failed to fructify in any appreciable measure, and to the extent that these were 'successful', the consequence was to force the labouring poor into the sub-urban areas to the east and south.[35] Officials were well aware of this exchange of population, though failing to note any irony in improving landscapes by displacing people. Beverley, the Census Commissioner of 1876, wrote: 'Every street that is widened, every new square opened, means so many persons displaced, and as the limits of the Town are fixed, many of them doubtless remove outside it altogether'.[36] A critique of the same was offered by the Scottish sociologist/planner Patrick Geddes who wrote of the 'dreary and conventional' plans that were unsparing of old homes and traditional neighbourhoods, the sweeping clearances being 'one of the most disastrous and pernicious blunders in the chequered history of sanitation'.[37] But rather than heeding to these critiques, officials transferred the blame to the cultural habits of the natives, especially the tradition of purdah which confined women and new born children to dark and ill-ventilated rooms with foul and foetid air. The absence of circulation of air inside thus mirrored the lack of ventilation outside, with many among Indians echoing similar sentiments about ingrained habits and the prospect of ill health:

> The description of a *kutcha* hut occupied by the poorer class of people with its number of inmates is something horrible like that of the Black Hole of Calcutta. People dread the admission of fresh air during the night, carefully closing windows and door, and they never forget to plug the small openings caused by broken panels of doors and windows with rugs and cotton. Those who have advised them to give up this suicidal practice of preventing access of air have incurred the censure of these foolish people.[38]

Things were not much different in Bombay either, even though it was more favourably located on the sea. The city, which had once been a forgotten island to which the East India Company had to make special efforts to draw people to trade, and it gradually eclipsed Surat as India's major port city by the nineteenth century. And as it grew beyond the original Fort area, it did so without taking care to 'lay down broad roads, provide open spaces for ventilation or to regulate the width of streets'.[39] Till the first decades of the twentieth century, most roads also continued to be ordinary water-bound constructions, with a surface crust of metal or sandstone but without any foundation. And even when rubble foundations were laid, tar introduced and the roads watered twice, much of the surface continued to be dry and dusty.[40] As for the interiors, among the major sanitary ills noted from the mid-nineteenth century was the familiar list of deficient sewerage, low and damp housing, overcrowding, defective ventilation, faulty water supply, ill-advised arrangement of dwelling houses, the presence of injurious trades, the putrefaction of intramural burial grounds, and the like.[41] Bombay had been once described as salubrious, and its sea breeze often evoked to suggest a convenient and healthy landscape, but its overcrowding was pronounced, affecting space and atmosphere alike:

> What has been said of Scotland may with equal truth be applied to Bombay, that families instead of living on the earth in the pure air with the sky above their dwellings, in many instances prefer lying stratum over stratum in flats opening into a common staircase, "a continuation of the street" as it has been called, which receives the organic emanations of the families on each floor.[42]

Fevers, 'peculiarly the disease generated by the foul air of crowded, uncleanly habitations', it was suggested, was prevalent far more in Bombay than in Calcutta or Madras, and the mass of the population had to breathe 'for ten or twelve hours out of the twenty-four the stifling atmosphere of their narrow and crowded chambers, loaded with animal exhalations'.[43]

Chawls and sheds made of dry leaves from date or coconut palms (*zavli* sheds) were the most common forms of housing for the urban poor, and the lack of regulations specifying minimum adequate living space, home amenities and open space between buildings implied that one room tenements were erected on 'every inch of available land', the primary motive being to extract maximum rent from ill-paid workers. Alongside, there was the expansion of the docks and growth of trade from the 1880s, drawing in a huge population, and in the process converting hut settlements in places such as Pathakwadi and Mandvi Kolivadi into 'the merest labyrinths of narrow and pestilential lanes', with 'crowded and sunless houses', the population huddled around 'poisonous tracks', sleeping amid 'the foulness of sewage' and 'breathing the miasmata that no breeze ever comes to disperse'.[44] By the late nineteenth century, Ira Klein points out, lower Colaba was 'thickly crowded and insanitary'; in Mandvi, buildings were piled almost atop each other and streets became 'narrow and overcrowded'; Nowroji Hill was 'ruined by its owner', who jammed together buildings without regard for 'breadth of streets and ventilation'; Chakla was 'full of dark and ill-ventilated milch-cattle stables'; and Chowpatty, Girgaum and other sections were 'full of cess pools' or filthy tanneries and 'horribly offensive' and 'unutterably foul' charnel houses, spurring 'environmental decay and disease'.[45] And as in Calcutta, so in Bombay, the health officer noted (1878) that in the absence of light and air in the buildings, 'the dissipation of the effete products of life is slow and incomplete, and they accumulate as poisons to destroy or blight the vitality of the human beings who live amidst them'.[46] In a concentrated form, such impure air could well destroy life; in dilution, they acted as poison, 'slower but no less certain', and to make matters worse, permitted or tolerated by law.[47] A new industrial city had come into being by the turn of the nineteenth century, but one marked by environmental degradation, human poverty and health pandemics.

It took the deadly plague of 1896 to stir things up, literally, when, in addition to the efforts at town sanitation, colonial authorities entered Indian homes to chase out the deadly rats and disinfect the native habitats. Humans, especially the labouring poor, and their habitations were both subject to extreme clean-up measures. Many were forcefully removed from their homes and confined in camps, disinfectants were poured into homes, earthen floors dug up, lime washing of houses done, obstructions to light removed, tillers taken off in many houses to let in air and light, and bamboo houses razed to the ground. As one officer put it, 'the dark, evil smelling, ill-ventilated, ill-drained, overcrowded lanes and alleys of Bombay were explored and thoroughly cleansed', though he forgot to add that the move was deeply resented both by the poor subject to it and the landlords who feared loss of rents.[48] An Improvement Trust was set up as a consequence in 1898, an outcome of a firmly entrenched belief that plague was, in the first instance, the direct result of overcrowding in poorly ventilated and filth-ridden dwellings.[49] The focus was twofold: an 'improvement scheme' to open up slums and 'street schemes' to draw in the breeze, from the ocean in the west into the neighbourhoods of central and east Bombay.[50] Moral reform combined with infrastructural aspirations, with Lord Sandhurst, the Governor of Bombay, desiring to place slum dwellers in better houses so that 'not only will it let the sun into their houses, but into their hearts and into their very existence and thus terminate the sad state of things amongst them'.[51] Immediate results were not encouraging though, with slums being demolished without alternative accommodation being provided for and too many rooms continuing to get too little light or air. 'Schemes for improving the City', the health officer observed, 'must necessarily press severely on the poorer classes'.[52] In time, a standard of light and air was suggested: 'a house cannot be built on a given area unless there is room in the space vertically above that area not only for the structure but also for the external open air space recognized by the 63.5° rule as indispensable adjunct to that structure.'[53] But not much success was to be had even then. At any rate, attention soon shifted away from a 'direct' attack on slums to an 'indirect' means of slum improvement, through a strategy of opening up areas in the north of the city where sanitary dwellings could be built to attract residents of the slums as also the business classes living in traditional congested areas.[54] Yet again, results were not too encouraging. If in Calcutta, this single-minded pursuit of ventilation as an antidote to a 'diseased' city 'was a declaration of war against the organic growth of Indian Calcutta itself', in Bombay, its immediate consequence was further overcrowding in the new development areas. Not surprisingly, the municipal corporation caustically remarked that these schemes of the Trust were anything but 'ventilation schemes' as claimed by the Trust, that is, street schemes designed either wholly or partly to remedy the defective ventilation of congested portions of the city.[55] As for the labouring poor, a full decade after the Trust had been working, their houses continued to be 'overcrowded, dark, damp and ill-ventilated', with the admission of fresh air into them 'a virtual impossibility'.[56] Not surprisingly therefore, attention again shifted to the habits of the people, in accounts of colonial officials and local elites alike, emphasising the 'natural' aversion of the poor to sanitation. The 'peculiar and perverse habits' of the people themselves, their 'ingrained prejudices and their want of personal cleanliness', wrote the physicians Jehangir J. Cursetji and Dinsha Bomanji Master, were responsible for the contamination of well waters, for flooding and for lack of ventilation, presenting a painful picture of a stolid and unconscious ignorance associated with great poverty, such as could be scarcely seen in the poorest civilised town of the West![57]

Factories: Dust and Humidity

Vitiated air was to be seen and felt not only inside homes or on the streets but also within factories. Despite the paucity of capital, the local business community in Bengal had embarked upon a broad range of steam-powered industries by the 1830s and 1840s, and for a time, it would appear, Calcutta was on the threshold of a small-scale industrial revolution.[58] Among the largest of the early enterprises were the manufactories based at Fort Gloster, about 15 miles from Calcutta, worked by steam for which coal was obtained from mines in Burdwan.[59] The city was not quite Manchester or Lancashire, not yet at least, but according to one account, there was much that was evident by way of coal use and industrial activity by the 1840s: 'on approaching Calcutta, the smoking chimneys of steam-engines are now seen in every direction, on either side of the river, presenting the gratifying appearance of a seat of numerous extensive manufactories, vying with many British cities'.[60] This was an even more likely scenario with the development of the jute industry after the 1850s, with mills being scattered on either side of the river Hooghly, north and south of Calcutta, and into the suburbs of Hooghly, Howrah and Sibpur.[61] Jute was not alone either, with observers mentioning iron works, jute presses, engineering works, cotton factories, flour mills and railway workshops among other industrial establishments in and around Calcutta.[62] In addition to these private factories, there were also new government and municipal establishments such as the mint and the water works. The town and the country were coming together, as it were, conjoined by mills and machines:

> Clean streets, garbage-less, lit by rows of gaslight,
> The full moon has come out, it's no longer night.
> Flour mills, jute mills, cloth and brick mills,
> Machines that dig out water and make landfills,
> Elephantine machines make a road a day,
> *Pranam* at the feet of machines,
> Town and country-have become twins.[63]

In Bombay, textiles took the place of jute, and the mills and rather than being dispersed, as in Calcutta, were concentrated in the very heart of the city. The initial pace of growth had been slow, with periods of prosperity precipitating frantic mill building, followed by slumps in which capital was swiftly diverted to other profitable uses.[64] Even then, by 1880, the city possessed 50 cotton mills, and the numbers steadily increased thereafter, going up to 83 by 1890.[65] 'Industrial capitalism', as one writer put it, 'drove out Civilianism who, therefore, planted itself thickly on the breezy Malabar and Cunballa Hills [while] the suburbs were appropriated by the cotton mills and the jerry builders of the day found chawls … for the accommodation of the growing industrial population.'[66] The textile industry, in turn, induced other works—engineering workshops, leather factories, dyeing and printing establishments, chemical works, and the like.[67] The mills also came accompanied by two major railway lines and three railway stations, along with workshops.[68] Accompanying these were other small establishments, in numbers so large that by the turn of the century, it had become more than evident that the only legislation that could be effective was the one that applied 'to the numerous small, ill-built, carelessly managed, and insufficiently equipped factories to be found in Bombay and other large towns'.[69] The consequence—a city not quite in the image of what may have been expected of *Urbs Prima in Indis*.

There was much that needed to be regulated about these industrial workspaces, with several interrelated factors pushing for specific interventions—from British precedence regarding legislation for working classes in England to development of an imperial public sphere desirous of humanist interventions, largely on account of pressure from the Lancashire lobby fearing competition of Bombay mills.[70] Of special concern were the hours of work for women and children, the contractual relation of adult male labour being conceived as entirely free and therefore outside the ambit of legislation. However, there were other conditions that affected all alike, pertaining to the presence of machines that might cause accidents and sanitary conditions that could lead to health epidemics. Questions regarding factory sanitation were thus variously addressed, enquiring into the state of water and air, quality of the ventilation, temperature and humidity inside work rooms, whether workers suffered from cotton dust and, if so, whether the establishment had any mechanical means for driving out the dust, etc. In April 1874, the coroner of Bombay and the health officer visited several mills, collecting information on sanitation and ventilation and reporting that a legislation was urgently called for 'to protect the lives, if not the health', of those employed in the cotton mills in the city.[71] A commission was consequently set up the year after, both to enquire into the conditions of work and the imperative for a legislation to regulate these.[72] After much deliberation, immediate legislation was ruled out, with the commission concluding that nothing ought to be done that might hamper the growth of an incipient industry.[73] However, the details on offer do give us an initial glimpse into the conditions that prevailed inside mills.

Divergent views were expressed regarding the use of windows, with some respondents insisting that mills used these only to let in sunlight, while others were of the view that windows sufficed both for light and for air. Many referred to cotton dust and workers having to cover their mouths with cloth, but while some suggested that workers did this out of choice, being 'lightly clad ... with nothing but a *langootee* on' and thus fearing chill, others were of the opinion that keeping windows and doors closed was a management decision so as to prevent draught and possible breaking of the thread on that count. The relationship between work conditions and ill health also elicited divergent views. The House Surgeon of Sir Jamsetjee Jejeebhoy Hospital linked chest diseases, inflammation of the lungs and derangement of the stomach to the atmospheric conditions of the mills, but the health officer only spoke of the dirtiness of the carding and blow rooms.[74] Most critically, members of the commission were not yet persuaded regarding the prejudicial effects of the atmosphere on the health of the workers.[75] It followed that the first Factory Act passed in 1881 remained silent on the issue, refraining from enquiring into the link between atmospheric conditions and health issues. And this was not its only limit. By defining a factory as an establishment that employed no less than 100 employees for at least four months a year, it left out of its ambit the numerous smaller workshops and factories in which labour was organised. On the other hand, by making this law applicable across the country, including in the jute mills of Bengal, it seems to have made Bombay interests somewhat happy![76] Calcutta, where there had developed a 'natural' unity between the jute industry and the local government, however, could not have been more unpleased, an unhappiness that it would continue to express in the post-legislative period.[77] In brief, the Act of 1881 seems to have disappointed many, both in India and in England.[78] Bombay would not sit content though, urging Government of India in 1884 to have another look at the Act and help remedy the defects. Indeed, the stage for this had been set some years earlier, by way of invitation to a factory inspector from England to help develop the necessary expertise for factory inspection in India. William Meade-King, a Scottish mid-level factory inspector from Lancashire, took up the appointment in 1882, followed after a brief interlude by James Jones, another trained inspector from Lancashire. Reports produced by these two, in turn, had a great bearing on the thinking in Bombay and London, respectively.

Meade-King's focus, like that of many before him, was also on minimum age, hours of work and, most critically, on arriving at a definition of factories that would include the majority of Indian establishments. But unlike some of the immediate enquiries that had preceded his, he also drew attention to 'the effluvia arising from latrines', 'dust and fluff flying about', 'acrid steam rising from sizing rooms', etc.[79] Even more was on offer in Jones's reports, especially concerning ventilation: 'This [accident], however, is not the greatest evil they have to put up with, for in the textile, bone crushing and flour mills ventilation and efficient means for carrying off the dust and gases arising from the various processes of manufacture are conspicuous by their absence.'[80] In the light of similar observations, Meade-King had proposed that a new clause be inserted in the Factory Act, stipulating that every factory should be kept in a clean state and should be ventilated in such a manner as to render harmless, so far as was practicable, all vapours, dust or other impurities generated in the course of the manufacturing process. And in instances where such conditions were not being met, it should be within the powers of the inspector to insist upon the use of mechanical devices to carry away dust and vapours.[81] No immediate change in legislation was initiated though, these reports merely helping to reopen questions concerning conditions of factory work in India and incidentally therefore also the question of air and ventilation, for the scrutiny of another commission.

The commission appointed to consider the working of factories in the Bombay Presidency in 1885 went over much the same ground as the commission of 1875, with opinion remaining divided on whether the new mills that had come up after the Act of 1881 were better built than the older ones. The use of steam, especially in winter months, was noted by nearly all who testified.[82] There was also near unanimity in considering the need for mechanical appliances such as fans in dealing with dust. The absence of ventilation in the factories engaged in corn grinding, wool cleaning, cotton cleaning and bone crushing, the filth of the bone dust, the stench and the swarming of flies that may not necessarily have been unhealthy but deserved to be abated, the crowding together of women workers into small and filthy unventilated sheds with low roofs, filled with dense dust in wool-cleaning factories, were some of the other issues to which attention was drawn afresh.[83]

The chief want of Bombay, the sanitary commissioner observed, was not light—in this respect India, and particularly Bombay, had an immense advantage over England—but an 'efficient system of ventilation and means for removing dust and fluff, especially in the smaller factories'.[84] Any legislative change that was envisaged in this regard, Bombay argued, must not apply to them alone but to all similar establishments across India. By contrast, respondents in Calcutta pointed out that conditions in the two provinces were entirely different, the managing agents of Cossipore Sugar Works going so far as to suggest that 'Bengal is essentially a foreign country when judged by the standards of Bombay'.[85] Given this difference, Government of India was unwilling to make matters compulsory, and even as a new Factory Act was legislated in 1891, on the matter of ventilation it was left entirely to local governments to make such rules as they might deem fit.[86] Calcutta, expectedly, saw no such need. Bombay, on the other hand, ruled that a factory would be considered ventilated only if openings were provided in the proportion of 10 sq. feet for each person in a room, admitting fresh air on a continuous basis.[87]

Expectedly though, this was not to the liking of the mill owners, who felt the rule was quite impractical. Indeed, the government itself seemed to have recognised the difficulty of putting it into practice. And as for factory inspectors, they had to remain content with reporting about such private initiatives as were being made regarding innovations in window design.[88] Here too, success was hard to come by. As the *Indian Textile Journal* put it:

> The subject of ventilation in cotton mills of India appears to have had less attention given to it than any other part of the business of manufacture. The millowners were indifferent to it because they seemed to be making their money without scientific study, and the authorities appointed by the Government to inspect factories have shirked it because they do not understand it.[89]

The *Times of India* offered its own strong, if paternalist, indictment, writing about the factory atmosphere the 'Bombay slaves' had to endure:

> The newer [mills] are generally well-lighted; there are parts of some of the older ones where the sunlight can never penetrate. The conditions of manufacture require a large amount of moisture at certain stages ... In most mills the windows seem to be perpetually closed, and they are covered with grime of years. I was assured that the work-people preferred it so and can quite believe that statement; I have often noticed the dislike of the industrial worker of India for fresh air. A similar trait is observable among the lower classes of England ... [but] *there are times when people must be protected even against themselves.*[90]

Enquiries regarding the possibility of introducing a standard of ventilation were finally started in the early years of the twentieth century. Departmental enquiries in England had already established that minimum cubic space per person was no guarantee of good ventilation, recommending therefore a standard specifying the proportion of carbon dioxide in the air.[91] India too could arrive at its own standards, it was believed, based on local tests, with the results being analysed by sanitary and factory experts.[92] Vitiated air, in other words, could and ought to move away from mere description to a definitive and measurable numerical expression, its contamination described on the basis of scientific tests, while keeping in mind the specific Indian context.[93] This required the aid of medical and chemical experts and it was hoped that these would be made available to the Textile Factories Labour Committee (TFLC) set up by Government of India in 1906. Questions relating to ventilation, however, were only briefly addressed by the TFLC, though it did make the important suggestion that temperature and humidity must also be considered alongside the issue of ventilation, bearing in mind both the efficiency of production and the health of the operatives.[94] The matter was addressed at greater length by the Indian Factory Labour Commission (IFLC), which followed the TFLC in October the next year regarding the state of affairs not only in Bombay but also in Calcutta. Responses from Bengal were brief and, as before, underlined the differences with Bombay. There was no humidification in jute mills, respondents suggested unanimously, making the question irrelevant. Similarly, the construction of factories was very different, being single storeyed with ample open spaces. The presence of a river also helped. As the chairman of the Indian Jute Mills Association put it:

> Jute mills are all built on the shed system, air space everywhere is very ample, and conditions generally conducive to the maintenance of good health … Jute mills are so well ventilated, and so comparatively free of dust, that I do not consider it necessary that samples of air should be taken in connection with the prescribing of an analytical standard for purity of air.[95]

The Government of Bengal endorsed the view, representing that it was 'at present impracticable to prescribe any analytical standard for purity of air in factories', the same remark also applicable as regards the standard of moisture. More crucially, they went on to observe: 'moreover, the fixing of such standards … would imply an elaboration of the system of inspection over factories, and an amount of interference with their working, which present conditions do not warrant.'[96] The 'natural' unity of government and capital in Bengal, as we observed earlier too, continued to hold true! Once again, therefore, the most detailed observations came from Bombay, but this time with the explicit recognition that in the absence of necessary health data, it was impossible to say whether mill hands suffered any more than other workers from ill health. Matters were not helped either by the fact that the outpatient records of the J. J. Hospital, situated in the mill district, were destroyed every three years.[97] All that was available was experiential evidence, and based on these, the health officer of Bombay opined that the mill hands of Bombay were, to all appearances, well-to-do, and with the exception of phthisis, suffered no more than the general population of the same class from any specific disease.[98] Other testimonies spoke more ambiguously, the joint medical inspector of factories, Bombay, suggesting, on the one hand, that operatives' health was not worse than those of others and, on the other, that 'from my medical knowledge I am inclined to believe that, where the heat and dust in the rooms are excessive, factory operatives must suffer from the effects of chills and irritation of the respiratory passage'. Yet others were categorical in asserting that the health of the mill workers was worse than those of others, with tuberculosis featuring more prominently among them:

> A very small percentage of persons suffering from phthisis, and working in a mill room, particularly an ill-ventilated one, are capable of giving the disease to other fellow-workers. They expectorate on the floors and the walls, and it is easy to conceive that an atmosphere laden with dust and cotton debris can form a very ready means of conveying about the tubercle bacillus.[99]

Not surprisingly, owners and agents of mills either elided the question of the ill health of workers altogether or side-stepped it through comparison with workers' housing, conditions of which were deemed inferior to those inside mills. But the commission seems to have been persuaded on this occasion, being of the view that mill operatives in Bombay did indeed suffer from ill health on account of long hours and 'confinement in an impure and humid atmosphere' and in 'often badly ventilated work-rooms'.[100] This was an important recognition, given past expressions of doubt on this count, but the fixing of standards would still be some years ahead.

Several inspectors had also been interviewed, who almost unanimously expressed the desirability of specifying standards of air purity, humidity and water purity, the only doubt being whether these should be based on English standards or as derived from local conditions.[101] Opinion among owners, by contrast, remained divided both as regard to the desirability of legislation and the extent to which it would improve matters. Workers' opinion on the issue, as before, was neither sought nor offered, the closest we come to hearing them being through the opinions expressed by Bhiwa Ramji Nare, a retired weaving master, who had worked in the industry for close to four decades and lived among mill hands:

> I don't know much about the analysis of air, but I think it will be difficult to prescribe a standard of purity which will be adapted to the different local and seasonal conditions, and to the work done in different departments of a factory. What is required is the maximum means of ventilation

> which would be availed of as to suit different conditions existing for the time being. The same is the case with a standard of moisture.[102]

Finally, it was for the Government of Bombay to offer its view, which was in the affirmative regarding the need for standards, but one that balanced the needs of manufacturing and the health of workers.[103] The commission's final recommendation were also positive, though partial. A standard of purity of air, they recommended, must be fixed so that a reasonable amount of ventilation may be insisted on. Owners of factories must also be persuaded to provide mechanical means, if necessary, to remove dust. However, given the variation in climatic conditions, a uniform standard of humidity was not quite called for![104]

Serious difficulties persisted, however, regarding the ability of factory inspectors to persuade mill owners to change their priorities.[105] The *Annual Factory Report* for 1908 mentioned that despite repeated instructions regarding the opening of windows in the weaving and spinning departments of the textile factories in Bombay, there had been 'little result'.[106] Next year again, the chief inspector of factories failed to 'report any progress in this important matter' and conditions in many factories remained 'far from satisfactory'.[107] A year later, there was 'no general improvement ... as yet', and to make matters worse, it was noticed that provision for inlet of fresh air made in the eaves of roofs in a few cases had been boarded up.[108] Some efforts were made, especially by 'some of the prominent mill owners' to improve ventilation, but matters of humidification and dust continued to go unaddressed. They also remained subject to economic calculations in that improvement was possible only to the extent that it did not lead to reduction in the breaking of the thread or the lowering of weight of the cotton.[109] Unfailingly, therefore, the demand of inspectors continued to be for a new legislation and the fixing of standards. Matters seem to have come to a head in October 1909, when the Textile Manufacturing Company Ltd was asked to remedy its ventilation within four weeks. The inspector had no such power to insist on remedies, the mill protested, observing that the rule of 1891, which insisted on provision of ventilation openings in the proportion of 10 sq. feet for each person, had been a dead letter from the very inception.[110] The matter was not of one factory alone, and the Bombay Millowners' Association (BMOA) also joined in the protest, resting content only after the collector's assurance that no rules regarding ventilation would be promulgated 'until due enquiry has been made and the proposed Rules referred to your Association for opinion'.[111] In this background, Government of India agreed to proceed on the lines recommended by the IFLC and in August 1911 recommended that the Bombay Government set up a Factory Ventilation Committee.[112] The committee duly undertook the task of visiting select textile factories throughout the Bombay Presidency—15 mills in Bombay, nine in Ahmedabad and two in Sholapur—and based on these enquiries suggested a CO_2 standard of:

In dry sheds ... 41 parts per 10,000;
In humid sheds ... 8 parts per 10,000.

These standards, they believed, were in conformity with the standards in England. True, but things had meanwhile changed in England itself, with recent research suggesting that the comfort of the operatives depended far more on temperature and humidity, thus casting doubts on a standard based on CO_2 alone.[113] These factors ought to be considered in India too, it was therefore suggested, and for this, assistance should be sought of a hygiene expert from England. The Secretary of State was duly approached for facilitating such a visit but in the meantime a war intervened, and any further enquiry into the question of ventilation got suspended.[114] In brief, over the course of the past half century vitiated air had gone from being an abstract concept to a definitive standard, with necessary instrumentation and protocols also recommended. But little would be operational till well into the 1930s and 1940s.

Smoke and Pollution

Homes, streets and factories, as sites for articulating the issue of ventilation, paralleled another concern around air purity, viz., the question of pollution on account of smoke, which increasingly came into view with the greater use of coal, in factories, homes and transport alike. Biomass did not entirely disappear, but dark black smoke came to be associated most closely with coal and, in time, demanded the greatest attention. But first, 'traditional' smoke.

Domestic Life and Other Quotidian Practices

Smoke found its first, though intermittent, mention in the bustee settlements, huts built of mud and straw, so low that their residents could hardly stand upright, and 'having no chimneys, the smoke of the fires with which they dress their victuals comes all out at the doors, and is perhaps more disagreeable to the passerby than themselves'.[115] This smoke, it is possible to infer, was still on account of the use of wood and dung, a situation that continued well into the nineteenth century, the magistrate of Howrah noting that the nuisance from the chimney smoke was 'infinitesimal' as compared with that from the cow dung cakes and other fuel used by natives. The commissioner of Burdwan similarly opined that 'people must cook their food and they can only use the fuel procurable. Wood fuel is scarce and cow dung, though objectionable, is the only fuel available for the poorer classes.'[116] However, a transition also seems to have been under way, with the special committee set up in 1879 to consider the smoke nuisance in Calcutta and its suburbs noting an increase in the volume of smoke on account of the substitution of charcoal, wood and cow dung by coal as fuel both for domestic and manufacturing purposes.[117] The sweetmeat-makers, wheelwright and coach builders, etc., the committee pointed out, increasingly used coal instead of charcoal or cow dung, and the trade in burnt coal for domestic purposes was steadily increasing. Not only was more coal being used, it was also of inferior quality, especially in comparison to English coal, containing a large percentage of sulphur which made the products of combustion 'very disagreeable to the senses.'[118] Fuel apart, there was also the issue of the lack of chimneys, of ill-ventilated kitchens full of acrid smoke and houses in which fuel burnt with a pungent odour:

> It [the fuel in native homes] is for the most part burnt within the dwelling and the smoke from it rises to the roof through which it percolates. The larger particles of soot are thereby deposited on the under-surface of the roof, and it would appear from this that, apart from the objectionable odour, there can be but little deposit of soot. Such smoke being situated near the ground level affects more particularly the inhabitants in the immediate vicinity of the dwellings.[119]

However, this was an issue that was to be really addressed at length only much later, in the 1920s, which we discuss in Chapter 5.

There were other aspects of the ordinary, quotidian life that were productive of smoke too and generated more immediate concern on this count.[120] Cremation amongst Hindus was one such issue, with questions being raised about its potential nuisance for those who lived in the vicinity of the ghats. The practice, though much disapproved by European observers, had remained rather unregulated till the middle of the nineteenth century, beginning when efforts were made to accommodate the practice within the evolving urban infrastructure and make it, as far as possible, compliant with sanitary expectations and environmental controls.[121] After much debate, a proposal was made in 1864 to move the burning ghat of Calcutta from Nimtollah to the Kasi Mitter Ghat in Bagbazar, on the view that the practice of openly burning the dead at a public ghat, in the midst of a crowded city, even though sanctioned by custom and by religious sentiment, was 'inconsistent with public decency and detrimental to public health'.[122] Not surprisingly, Indian members of the Justices of Peace were reluctant to endorse such a move, arguing that not only would this be an inconvenience to members of the native community but also that there was no evidence that the natives or Europeans perceived the ghat as a nuisance.[123] They were willing, however, to consider ways that would make the burning of bodies as inoffensive as possible, in part by preventing exposure of the general masses to the burning and in part by ensuring that bodies were burnt in such a manner as to prevent the escape of the gases except through a lofty chimney. Various designs were proposed and efforts were made to obtain additional land, but money remained scarce. Special incinerators designed by the inspector also remained unsatisfactory and soon dismantled in 1868.[124] And while prevailing medical opinion was convinced that the practice of open cremation was a nuisance, it was not quite certain whether it could also be condemned 'as noxious to life, or even health' or whether the fumes and effluvia they produced necessarily gave rise to diseases:

> The whole subject of the causation of disease by miasmata of this and similar character is too much involved in obscurity to justify dogmatic assertion that such is the case. But of this there can be no doubt, that the contamination of the atmosphere with foreign and noxious effluvia, such as those arising from decomposing or burning animal matters, though it may not effect individuals employed ... or small sections of the community living in the vicinity, yet on the whole is prejudicial to public health and therefore comes clearly under that class of nuisance which should be excluded from the interior of cities.[125]

This is where matters stood for a long time, though not without provoking some resentment at the altered 'deathscape':

> The city is now unlivable.
> Such, mother, is our present plight:
> Even death offers no respite.
> None any more will light our pyres
> We'll now be burnt in incinerators.[126]

It was not the practice in itself but the manner in which it was carried out that was objectionable, argued the health officer. And in all such matters that concerned religious sentiments, on the one hand, and health and comfort of the public, on the other, a necessary balance must be found, opined the Calcutta High Court.[127] In the years to come, cremation practices came in for some appreciation too, even in Western countries, this being considered more hygienic than traditional burial, so long as the smoke and gases were controlled for. Lending support to this view in India were influential bodies such as the *Indian Medical*

Gazette, which suggested that under such climatic conditions where putrefaction was rapid and the risk of disease transmission much greater, cremation was by far the more desirable practice from a sanitary and environmental perspective.[128] However, the appreciation often came together with a rider that what ought to be promoted was not the open-air cremation preferred by upper caste Hindus but indoor cremation in purpose-built structures, using gas or electric burners that generated very high temperatures and rapidly reduced a corpse to ashes. A crematorium of this kind was built in Calcutta in 1903, based on the recommendation of the Cremation Society of Bengal, with the heat of the furnace argued to be sufficient to consume all smoke and gases. But the practice remained rather limited, with only four cremations reported to have taken place in 1913–14.[129] In Bombay too, the open burning ghats had elicited adverse remarks from the British in the mid-nineteenth century:

> To ride home to Malabar Hills along the sands of Back Bay was to encounter sights and odours too horrible to describe—to leap over sewers whose gaping mouths discharged deep black streams across your path, to be impeded as you neared Chowpatty by boats and nets and stacks of firewood, and to be choked by the fumes from the open burning ghat.[130]

However, as in Calcutta, a more favourbale perspective towards cremation became evident by the closing years of the nineteenth century, especially in the context of the plague of 1896.[131] Notwithstanding this shift, important differences remained, with the Hindu upper caste unwilling to accept the modern Western practice of indoor cremation as being ritually appropriate, leading David Arnold to surmise that 'if in Britain cremation progressed down the social scale from the upper and middle classes, in India Western-style cremation began with the low-caste poor and struggled to gain acceptance among high-status communities'.[132]

Closely aligned to the question of cremation was the issue of the disposal of rubbish by incineration.[133] In Bombay, 883 tonnes of rubbish was being handled daily by the turn of the century, the corresponding figure for Calcutta being 1,000–1,500 tonnes, a task beyond the capacity of traditional methods of handling waste through their consignment in water bodies (rivers and seas) or on landfill sites. One response was to make the land and water disposal systems more efficient; the alternative was to burn it, though the question remained whether in doing so, one form of pollution (on account of solid waste) was being exchanged for another (smoke and air pollution). As Henry Conybeare put it in his report on sanitation in Bombay (1855), not only would the burning of night soil be costly, necessitating the use of large volumes of coal (or alternatively wood), but it would also be a nuisance:

> the liquid portion of the night-soil, forming 75 percent of its whole bulk, would be expanded 1800 times by the process of vaporization, and from the way in which smoke and vapour often appear to descend and rest on the surface of the low parts of the town during the early morning, I believe, however high a chimney were provided, the steam of the boiling ordure would occasion serious nuisance in certain states of the atmosphere, which are by no means of rare occurrence.[134]

Notwithstanding, the city experimented with a number of 'crude incinerators' before the municipal engineer decided to install a more sophisticated British destructor. But even with this, success was moderate, the stench of the assembled garbage, the smoke from the chimney and the costs involved leading to its quick abandonment. In 1895, the question was considered anew, especially the possibility of many small incinerators located in different neighbourhoods substituting for a large centralised one. The risk of smoke nuisance over the island, however, remained. Raising the height of chimneys could be a possible remedy to this nuisance, but this would enhance the cost of construction. And so, both on grounds of nuisance and because of the costs involved, it was argued, incinerators were not a good idea, especially not seven or eight of them located within the city limits![135]

Much the same was the story in Calcutta where proposals for introducing incinerators were first made in 1863, borrowing from a plan that had been successful in Karachi. The experiment was short-lived here too, though revived again in the 1880s. Once again objections were immediately raised. Unlike European and American waste that contained ashes in the form of breeze that kept up the fire in the furnace without needing extra fuel, the waste of Calcutta, it was pointed out, contained no ashes and would require much additional coal to burn. In addition, the waste of Eastern cities contained much greater moisture content, requiring additional fuel to first dry them before burning. However, the health officer was not persuaded, arguing that the differences in qualities of waste notwithstanding, all waste was fit for combustion, though the incinerator may have to be designed differently in India than in England. The *Statesman* too wrote in his support, privileging incineration over 'the present disagreeable practice of shooting the refuse down at the salt marshes, there to rot and pollute the air'.[136] An experimental incinerator was subsequently started at Entally and was quickly declared a success, even though observers noted that the products of combustion proceeding from the chimney were 'apt at times to be offensive'. The addition of special fume cremators, it was believed, would provide a solution to this and particular mention was made of the Harrington incinerator, capable of burning all offensive gases, requiring little additional coal and being easy to work even by unskilled stokers. But perhaps success was anticipated too quickly regarding this too. Ironically, an early warning came from the *Statesman* itself, one-time enthusiast of Harrington's design, but now recognising that an expansion of this mode of disposal would require for the incinerator to be placed close to the refuse dump on Circular Road where it would be surrounded by huts and houses. 'It will be no gain to the health of the town', they opined, 'to poison people living within a certain radius of each incinerator by means of smoke from incinerator chimneys. The objection to the smoke from the incinerator is precisely the objection that was taken to incinerators of the Harrington type in England.'[137] The *Indian Medical Gazette* that had welcomed trials for burning Calcutta's waste in 1889, merely six years later, was also far from enthusiastic about the same, citing adversely both the smell and smoke produced and the high costs involved.[138] Even Dr Simpson, the ardent advocate of incineration in Calcutta, eventually ceased to recommend it for crowded or important localities, though the vague promise continued to be entertained that if incineration was done someplace outside the town, many of the nuisances complained of could be avoided.[139] And so, Arnold writes, apart from the growing size of the urban rubbish mountain, little seems to have seemingly changed after the 1910s![140]

Steam Engines and Smoky Skies

Domestic fires, burning ghats and incinerators apart, it was factory smoke that was to occasion the greatest concern. By the turn of the century, there was much about Calcutta that was celebrated, especially by Englishmen who saw the city as primarily their creation, the capital of not merely a province but of the Indian Empire. Its wonders drew visitors from far and near to 'see the river with its crowded shipping', 'the quays with the jostle and clamour of their busy life', 'the Howrah bridge, so useful and so inadequate', 'the jute-mills and cotton mills drawing their sooty finger-marks across the sky', 'the Government buildings and the law courts', 'the glorious and health giving expanse of the Maidan', etc.[141] The poet Ghalib who visited the city in 1828 could still recall with wonder some 20 years later the changes under way on account of steam power:

> Heavens, what a magic have they breathed in the breast of the sea that it lets the smoke pilot ship after ship on its waters. And what a wonder! The smoke steers the ship and carries it once as far as the Oxus, while still other times brings the world to the very sands of a desert. The boat receives its speed from the smoke, and both the wind and the wave come to fight with it.[142]

Given Calcutta's geography there was much to be seen and felt about this smoke too. Rudyard Kipling described the uncanny similarity this perception produced between Calcutta and 'home':

> Let us take off our hats to Calcutta, the many-sided, the smoky, the magnificent, as we drive in over the Hughli

> Bridge in the dawn of a still February morning. We have left India behind us at Howrah Station, and now we enter foreign parts. No, not wholly foreign. Say rather too familiar ... Calcutta holds out false hopes of some return. The dense smoke hangs low, in the chill of the morning, over an ocean of roofs, and, as the city wakes, there goes up to the smoke a deep, full-throated boom of life and motion and humanity. For this reason does he who sees Calcutta for the first time hang joyously out of the *ticca gharri* and sniff the smoke, and turn his face toward the tumult, saying: "This is, at last, some portion of my heritage returned to me. This is a city. There is life here, and there should be all manner of pleasant things for the having, across the river and under the smoke."[143]

Kipling's 'almost home' was, however, other people's despair. The absence of town planning implied that industry had been set up rather indiscriminately, and often in the immediate vicinity of residential areas.[144] By the turn of the century, there was much that would be disliked about such intermingling of mills and residence, especially concerning the pall of smoke that seemed to spread over neighbourhoods.[145] K. C. Bose characterised it as one of the 'auxiliary defects' of the city and 'an evil of no small magnitude', to be forthwith removed.[146] Babu Nalin Behari Sircar testified to the great annoyance felt by residents of the city, especially in Wards 3 and 4, while Babu Ambika Charan Mathur characterised it as a stigma to the 'City of Palaces', going so far as to suggest that the one who could get rid of Calcutta's smoke was worthy of a Nobel Prize as the one who might find a cure for cancer![147] Humans apart, there was much to be feared for nature too—Durgacharan Ray's Bengali text *Debganer Marttye Agaman* imagined Ganga complaining to the Gods who had arrived in Calcutta in 1880 about the indignities suffered by her at the hands of the English, not least on account of the noise and smoke from the land that had been snatched from her to put up the massive factories![148] For all the allusion to industrial progress and urban growth, there was not much love for smoke in Calcutta, both among bhadralok Bengalis and among the English administrative elite. It was not only about what had come to be but also what could be anticipated as the (blighted) urban future. Viceroy Curzon gave expression to this in no uncertain terms in his address to the Bengal Chamber of Commerce (1903):

> But there is one superficial feature of Calcutta that has greatly distressed me ... I allude to the Calcutta smoke, which sometimes almost makes one forget that this is an Asiatic capital, which besmirches the midday sky with its vulgar tar-brush and turns our sunsets into a murky gloom. I am reluctant to see Calcutta, which has risen like a flame, perish in soot and smoke.[149]

In much the same way, as industrial presence in Bombay increased, so did the perception of smoky skies. Indeed, even before the city had become India's industrial capital, its residents were beginning to be acquainted with the dystopian future that awaited them. Smoke, informed the *Bombay Times and Journal of Commerce* in March 1840, 'is always associated with recollections of nature in a state of pollution, and of feeling oppressed, tortured and stupefied by all that is odious and unsightly'.[150] Smoke to look upon, smoke to breathe, smoke to eat, drink and sleep amongst, unending smoke day and night, such had become the fate of Glasgow, Manchester, Liverpool, Birmingham and London. Would this also be Bombay's future? The *Times of India* editorialised about economic waste on account of smoke, the dangers that it posited and the discomfort that it produced. Memorialists wrote of the dirt felt within homes and on streets alike and, in time, health hazards figured prominently too. And at all times, there was the potential nuisance to be countered, especially with the increasing use of cheap, country coal:

> Day after day they pollute the only respectable road in Bombay with dense clouds of filthy smoke. The merchant rolls down to his office with a smut in his eye, and his temper is bad for the rest of the day. The fashionable lady, who sets out so spotless from the Ridge, reaches the Yacht Club a grimy, smoky bundle. The sea, that should look blue, is a dull grey. And all this is the fault of the B. B. & C. I.

> They are burning bad coal, cheap and nasty, and we have to pay for it.[151]

Private spaces were as much affected as the public streets, residents of Duncan Road memorialising the Bombay town council in 1880 about the dingy appearance of their homes on account of smoke and about the dust that came from the wool-cleaning factories—dust on their furniture, dust on their clothes, papers and books, such that wherever they turned, they 'seemed to be moving in poisonous atmosphere of dust'.[152] Similarly, the *Times of India*:

> in the neighbourhoods of the cotton mills the buildings are becoming as grimy as they are in Bolton or Preston, and in a situation where the best private bungalows in the island are built, all the inhabitants are grumbling at the clouds of smoke and smuts that render the early morning intolerable.[153]

Rents suffered too, as alleged by owners and residents of the Grant Building, on account of the smoke, dust and cotton fluff that issued from the adjoining Nicol Mill.[154] Like Kipling's account of Calcutta, there was much to be compared with smoke back home and like Curzon's observation, there was much that would be considered undesirable in such comparison:

> No great town in the world is more beautifully situated or more picturesquely surrounded, but human ingenuity is doing what it can to destroy all the advantages so lavishly bestowed by Nature. What with the stenches, innumerable as those of Cologne, from the city drains below, the clouds of smuts in the smoky atmosphere above, and the unsavoury breezes that are wafted in from the rising tides on all sides, one might as well live by the North Woolwich Outfall or at Leeds or Glasgow, as in this finest of oriental capitals.[155]

The debate concerned railways and shipping as much as it did factories. The use of cheap coals by the railways, it was suggested, was because the government was desirous of encouraging the industry by lowering costs—to use English coal instead would cost almost a lakh rupees extra each year, unacceptable to shareholders and government alike.[156] In turn, the *Times of India* took the defenders of the economic logic to task:

> They appear to assume that the public are compelled to bow before the sacred altar of dividends, and that thousands ought to submit to a daily infliction that has grown intolerable, if the remedy involves any additional expenditure. It appears to be necessary to remind these gentlemen of limited views that the comfort and convenience of the public must take precedence even of the greed for profits; and that, it is of greater importance that the people of Bombay should not endure an unnecessary nuisance, and that the main thoroughfare of the city should not be ruined as a pleasure resort, than that avaricious shareholders should pocket a few thousand extra rupees.[157]

In time, other issues got conjoined, in Calcutta and in Bombay, ranging from smoke causing 'distemper of diverse kinds', to lowering vitality, reducing the fertility of the soil, poisoning the food supply, corroding masonry buildings, blackening household linen, cutting off the sun's health, influencing atmospheric conditions by creating fog and gloom, and creating conditions detrimental to the physiology of the population.[158] This was smoke that impacted all alike, Europeans and Indians, old and young, rich and poor (see Figure 1.1).[159]

The damage was seen as being not limited to smoky cities alone but extending far into the surrounding country, with corroding gases being carried along with the wind.[160] Gender came to the fore as much as race. Polluted air, sickness of family members, household goods destroyed by grime and dust, the never-ending wear and tear, and the worry of cleaning, all of these added to the burdens of women, observed the Bengal Smoke Nuisances Commission in 1930. And while these may have been the experience of people of all classes, the poor, lacking even basic amenities, were the greats sufferers.[161] The point was underlined in subsequent reports too, the commission observing that not only was injury caused on account of carbon in the air but also it was common practice for women to shut the windows of their homes to prevent smoke and smuts from entering, thereby aggravating

Calcutta Smoke Ruins Cold Season.

These are Calcutta's weekly victims poisoned by polluted air, when the aerial scavengers wind and rain fail and the air we breathe becomes thickened with impurities, and the nose and respiratory tract are clogged with soot, and household linen and clothing are blackened in one day.

Source: Bengal Smoke Nuisances Commission, *Annual Report*, 1928, p. 2.

Figure 1.1: Diverse Victims of Calcutta Smoke

the nuisance.[162] However, even as women and the poor were to elicit sympathy as *residents*, as *workers*, there was an opposite perspective on offer, with the responsibility of keeping the smoke within stipulated time limits shifting by the 1910s from European managers to Indian stokers accused of 'careless' firing.[163] Prejudice against native firemen loomed large. Grover, in his report on Calcutta referred to the 'low standard of intelligence of the native fireman', to their 'stolid indifference', to their 'passive resistance', and only then to the onerous conditions of heat in which they worked, which made it natural for them to get through their work in the shortest possible time which they did by putting on the fire a maximum quantity of coal and retiring to a cool place.[164] And when a Smoke Nuisances Commission (SNC) came to be established in Bombay in 1912, they too were of the view that it was the stokers who were most responsible for excessive black smoke, though tempering their remarks with the observation that firing once in 20 minutes rather than three times was less work, and almost excusable given the use of inferior coal, excessive heat from the boilers and excessive natural heat in Indian conditions. Race and class, thus, would be as much implicated in the discourse of air, as in other strategies of the making of the urban modern in India, but perhaps in a more subtle manner, contaminated air being argued to affect all alike.

One way of dealing with this contamination and smoke could have been through the relocation of those establishments that produced excessive smoke. Thus, for instance, there were suggestions to move jute mills out of Calcutta in 1860 and to move industries to the north-east corner of Bombay in 1914.[165] However, the strategy had its limits, for not only would the workers who were part of these establishments and lived near it continue to suffer from smoke, given favourable wind conditions, it was likely that smoke and soot would drift in from these locations back into the city.[166] This left the option of regulation, of reducing smoke at source, but even as this was so, it was not always evident just what the issue at stake was, of what *really* mattered. Colonial officials, though keen on legislating against smoke, remained fuzzy on how best to 'frame' it. In 1905, for instance, when a new Smoke Nuisances Act was framed for Bengal, it was popularly believed to have been enacted to protect the white marble of the Victoria Memorial![167] The Home Department reasoned that the legislation had been undertaken because of smoke being a *nuisance*, and not merely on account of objections from a sanitary perspective.[168] H. H. Risley, secretary to the Government of India, concurred. There was little denying that there would be some sanitary interests, he wrote, but the smoke nuisance in Calcutta was far more intimately connected with commercial interests and factory questions, such as the quality of coal, boiler power, method of stoking, etc. The working of the Smoke Nuisances Act, he opined, was rather like the working of the Factories Act, administered by the Department of Commerce and Industry, and it was therefore logical that the responsibility for framing rules and execution must rest with them.[169] The Department of Commerce and Industry was hardly amused at the prospect of this additional responsibility. The matter of smoke nuisance, they argued instead, was primarily in the interests of the health and comfort of the citizens of Calcutta, which concerned their department only incidentally.[170] The other concern was aesthetic, and none could contend that the

Department of Commerce had any deep interest in that question! They, in turn, were backed by officials who argued that Government of India was hardly likely to have interfered with the firing of furnaces with a view to reducing the cost of running them. That intervention needed a public purpose, which in this view could only be 'to prevent injury to the appearance of Calcutta and to the health of its population'.[171] In this regard, laws and policies regarding smoke in colonial India may not have been quite different from interventions in the metropolitan contexts, the struggle against smoke in England also being argued to be simultaneously about several things: the attempt to control processes by which 'the glorious sunlight is so much and so often shut out from lighting the street and byways of our great city', 'fighting all that militates against the health of the individual and the community' and ensuring that there was little economic loss, which would be the case if carbon burned to carbon monoxide, producing smoke that would give only a third of the heating power than if the carbon burned to carbon dioxide.[172] At another level though, smoke in Bombay and Calcutta, or indeed in other cities such as Karachi, Kanpur and Ahmedabad, was perceived as being 'unnatural', as an uncanny and undesirable presence in Indian contexts. And so, while Bengalis continued to ignore dung smoke, and the English romanticised wood smoke, both were frequent in complaining about the smoke from coal-fired mills.[173]

There was much to be gained from metropolitan expertise, as also important lines of distinction as regards regulating this smoke.[174] Unlike in England, where smoke nuisance seems to have been integral to the sanitary discourse right through, in India, the difference between the two became more pronounced over time with the task of regulating smoke being taken away from municipal authorities, after the setting up of the Smoke Nuisances Commissions in Calcutta and Bombay.[175] Anti-smoke zactivism in Europe and the United States had, among other things, placed smoke as a moral concern, leading to stunted lives, especially among the working classes. In India, by contrast, while anti-smoke commissions would share the economic, aesthetic and health arguments with their European peers, only occasional mention was made about smokiness and darkness as inducements to crime and immoral behaviour, lethargy and mental degeneration.[176] Prominent civic associations in England included the Ladies' Sanitary Association, the Manchester and Salford Noxious Vapours Abatement Association, and the Smoke Abatement League. Their USA counterparts included the Smoke and Dust Abatement League of Pittsburgh; the Smoke Abatement Leagues of Chicago, Cincinnati and Baltimore; the Women's Smoke Abatement League of Chicago and later of St Louis; Women's Health Protective Association; and the Woman's Club of Cincinnati.[177] In India, once again by contrast, civic associations were entirely missing when it came to the issue of smoke.[178] To be certain, protesting voices were not absent, but they registered themselves largely as individual complaints, either in courts or in newspaper columns.[179] And even when claims were made on behalf of specific groups—as, for instance, the hardships faced by housewives invoked with some frequency from the 1920s—this was done not through a civic group but through the Smoke Nuisances Commissions on which sat representatives of commerce, industry and government alone. Florence Nightingale had once famously observed that in India the government was 'everything', having to achieve single-handedly what had been done in Britain in conjunction with 'habits of self-government'.[180] The story of smoke would suggest that in this respect at least Nightingale's observation was entirely right, with the proviso that much of what the government would take credit for would be done in collaboration with industry, especially European mill managers, whom officials often sought to inform and advise, and occasionally prosecute. Finally, there were also the peculiarities of the Indian conditions to be considered—regarding the quality of resources, the character of the labour force, in matters of expertise and governmental powers, and as regards domestic architecture and the burdens of labour inside homes. In brief, the promise that European science and technology held out and which native materials, practices and institutional arrangements seemingly undermined.

It is this history of smoke and its regulation in Indian cities that I narrate in the chapters ahead.

Notes

1. François Bernier, *Travels in the Mogul Empire, 1656–1668*, translated and annotated by Archibald Constable, 1891, 2nd edition, revised by Vincent Smith, London: OUP, 1916, p. 351.
2. Bernier, *Travels*, p. 368.
3. 'Bernier's Travels in the Mogul Empire', *The Oriental Herald and Colonial Review*, 10, 1826, p. 468.
4. For a similar discourse of stench and modernity in China, see Xuelei Huang, 'Deodorizing China: Odour, Ordure, and Colonial (Dis)order in Shanghai, 1840s–1940s', *Modern Asian Studies* (hereafter *MAS*), 50, 3, 2016, pp. 1092–1122.
5. I have found it especially challenging to specify such distinctiveness through a history of smoke and air pollution as almost all available historical literature is about European and American cities. See Stephen Mosley, *The Chimney of the World: A History of Smoke Pollution in Victorian and Edwardian Manchester*, Abindon, Oxon: Routledge, 2008; Peter Thorsheim, *Inventing Pollution: Coal, Smoke and Culture in Britain since 1800*, Athens, USA: Ohio University Press, 2006; Frank Uekötter, *The Age of Smoke: Environmental Policy in Germany and the United States, 1800–1970*, Pittsburgh: University of Pittsburgh Press, 2009; David Stradling, *Smokestacks and Progressives: Environmentalists, Engineers and Air Quality in America, 1881–1951*, Baltimore: Johns Hopkins University Press, 1999. In India, by contrast, we have a solitary pioneering essay by M. R. Anderson on smoke in Calcutta published more than two decades back! See M. R. Anderson, 'The Conquest of Smoke: Legislation and Pollution in Colonial India', in David Arnold and Ramchandra Guha, eds, *Nature, Culture and Imperialism: Essays on Environmental History of South Asia*, Delhi: OUP, 1995, pp. 293–335.
6. The standard binary being of 'wide fields and fresh air of the village', on the one hand, and 'cramped and often insanitary streets and lanes of the town', on the other. See Royal Commission on Labour in India, *Evidence*, Vol. I, Part 1, Bombay Presidency (Including Sind), London: His Majesty's Stationery Office, 1931, p. 17.
7. For a creative interpretation of the relation of cities and villages in India, see Ashis Nandy, *An Ambiguous Journey to the City: The Village and Other Odd Ruins of the Self in the Indian Imagination*, Delhi: OUP, 2007.

8. See Nik Heynen, Maria Kaika and Erik Sywngedouw, eds, *In the Nature of Cities: Urban Political Ecology and the Politics of Urban Metabolism*, London and New York: Routledge, 2006; Lisa Benton Short and John Rennie Short, *Cities and Nature*, Abingdon, Oxon: Routledge, 2008; Ian Douglas, *Cities: An Environmental History*, London: I. B. Tauris, 2013. In a pioneering essay written a decade or so before these works, Christine Rosen and Joel Tarr had grouped urban environmental history under four broad rubrics: (i) analysis of the effect of cities on the natural environment, (ii) analysis of the impact of natural environment on cities, (iii) the study of societal responses to these impacts and (iv) examination of the built environment and its role and place in human life as part of the physical context in which society evolves. See Christine Rosen and Joel Tarr, 'The Importance of an Urban Perspective in Environmental History', *Journal of Urban History*, 20, 3, 1994, pp. 299–310.
9. Maureen Flanagan, 'Environmental Justice in the City: A Theme for Urban Environmental History', *Environmental History*, 5, 2, 2000, pp. 159–64.
10. See Ishita Pande, *Medicine, Race and Liberalism in British Bengal: Symptoms of Empire*, London: Routledge, 2010, p. 98. For a study of hygienic discourse as being not only about cleanliness but also about becoming modern, see Ruth Rogaski, *Hygienic Modernity: Meanings of Health and Disease in Treaty-port China*, Berkeley: University of California Press, 2004.
11. James Renald Martin, *Notes on the Medical Topography of Calcutta*, Calcutta: Bengal Military Orphan Press, 1837, p. 96. Christopher Hamlin suggests that in India environmental determination remained strong for much of the nineteenth century, unlike in Europe, where clinical research and laboratory-based microbiology began to eclipse atmospheric determinants of health after the dis-appearance of malaria. See Christopher Hamlin, 'Surgeon Reginald Orton and the Pathology of Deadly Air: The Contest for Context in Environmental Health', in James Rodger Fleming and Ann Johnson, eds, *Toxic Airs: Body, Place and Planet in Historical Perspective*, Pittsburgh: University of Pittsburgh Press, 2014, pp. 23–49.
12. W. J. R. Simpson, *The Principles of Hygiene as Applied to Tropical and Sub-tropical Climates*, London: John Bale Sons & Danielsson, 1908, p. 294.
13. It ought to be mentioned though that the discourse around smoke/nuisance/waste/pollution was always a marginal presence within the

larger discourse on filth and cleanliness. This was true of things such as noise and smell too, which found only episodic mention in urban discourses, both in India and elsewhere. See Melanie Kiechle, *Smell Detectives: An Olfactory History of Nineteenth Century Urban America*, Seattle: University of Washington Press, 2017.

14. Value could be attached to sanitary waste too, as, for instance, in the use and sale of human excreta as fertiliser. See Paul Kreitman, 'Attacked by Excrement: The Political Ecology of Shit in Wartime and Postwar Tokyo', *Environmental History*, 23, 2, 2018, pp. 342–66.
15. Timothy Cooper makes an analytical distinction between filth and waste, the first being about abjection and the latter about practices of valorisation, which he considers to be a classically modern European impulse. I prefer instead to think through the dual nature of waste itself, considering it simultaneously as a material thing (which could be productive of pollution) and a normative principle (regarding wasteful and efficient use of resources). See Timothy Cooper, 'Recycling Modernity: Waste and Environmental History', *History Compass*, 8, 9, 2010, pp. 1114–25 and 'Modernity and the Politics of Waste in Britain', in Sverker Sorlin and Paul Warde, eds, *Nature's End: History and the Environment*, London: Palgrave, 2009, pp. 247–72.
16. Damage on account of smoke and wasteful use of fuel thus provided two different economic rationales for regulating smoke. It was more the latter meaning—smoke as inefficient use of fuel—that was foregrounded in India. By contrast, in places where coal was quite cheap, it was the damages caused to property that merited more attention. See Angela Gugliotta, '"Hell with the Lid Taken Off": A Cultural History of Air Pollution—Pittsburgh', Ph.D. diss., Notre Dame University, 2004, pp. 390–92, 454–55; Richard Hawes, 'The Municipal Regulation of Smoke Pollution in Liverpool, 1853–1866', *Environment and History*, 4, 1, 1998, pp. 75–90.
17. So just as the question of value was not entirely absent from considerations of sanitary waste (see note 14), the matter of disposal (through dispersal) also figured in considerations of the loss of value on account of partially burnt coal.
18. For sanitary networks in Bombay, see Mariam Dossal, *Imperial Designs and Indian Realities: The Planning of Bombay City, 1845–1875*, New York: OUP, 1991; Mridula Ramanna, *Western Medicine*

and Public Health in Colonial Bombay, 1845–1895, Delhi: Orient Longman, 2002; Prashant Kidambi, *The Making of an Indian Metropolis: Colonial Governance and Public Culture in Bombay, 1890–1920*, Burlington, Vermont: Ashgate Publishing Company, 2007; Sandip Hazareesingh, *The Colonial City and the Challenge of Modernity: Urban Hegemonies and Civic Contestations in Bombay City, 1900–1925*, Delhi: Orient Longman, 2007. For Calcutta, see Mark Harrison, *Public Health in British India: Anglo-Indian Preventive Medicine, 1859–1914*, Cambridge: Cambridge University Press, 1994; Pande, *Medicine, Race and Liberalism*; Partho Datta, *Planning the City: Urbanization and Reform in Calcutta, c.1800–c.1940*, Delhi: Tulika Books, 2012.

19. Miasma, it was believed, could be generated both by decomposing waste matter and by the breath and sweat that accumulated in overcrowded interiors, especially when the air was stagnant.
20. On views of sanitary reformers regarding the English working class, see Michelle Allen, *Cleansing the City: Sanitary Geographies in Victorian London*, Athens: Ohio University Press, 2008.
21. On colonialism as a civilising mission, see the collection of essays included in Harald Fischer-Tiné and Michael Mann, eds, *Colonialism as Civilising Mission: Cultural Ideology in British India*, London: Anthem Press, 2004. Characteristic of this was the desire for technological progress that remained socially conservative.
22. Captain Hamilton, who has been referred to as the 'Sindbad of the eighteenth century', traded and travelled by sea and land between the Cape and India and Japan, between 1688 and 1723. This particular comment by him has been much cited in subsequent histories of Calcutta. I borrow from H. E. A. Cotton, *Calcutta Old and New: A Historical and Descriptive Handbook to the City*, Calcutta: W. Newman & Co., 1907, p. 10.
23. Martin, *Medical Topography*, pp. 14–15; *Calcutta Gazette*, 19 Oct. 1786, cited in A. K. Ray, *A Short History of Calcutta Town and Suburbs*, Census of 1901, Vol. VII, Part I, 1902, reprint, Calcutta: Riddhi India, 1982, p. 151.
24. Martin, *Medical Topography*, p. 23.
25. The Lottery Committee was set up to initiate improvement schemes for Calcutta for which money was to be raised through public lotteries. Hence the name.
26. Cited in Datta, *Planning the City*, p. 38.

27. For the theory of the free flow of air as constituting a sanitary street paradigm in England, see Michael Hebbert, 'A City in Good Shape: Town Planning and Public Health', *Town Planning Review*, 70, 4, 1999, pp. 433–53.
28. 'Advantages of the Use of Gas in Private Houses in Calcutta with a Description of the Manufacture of Coal Gas', Calcutta: Thacker, Spink & Co., 1854, p. 4. More than half a century later, addressing the first All-India Sanitary Conference in 1911, K. C. Bose would similarly observe that the dust of the streets of Calcutta was 'impregnated with the germs of Phthisis' and 'when wafted by the wind' got access into the respiratory systems of the people, while the narrow streets and lofty houses obstructed the 'perflation of air and the access of the sun'. See K. C. Bose, *Tuberculosis*, Appendix 21, Proceedings of the First All-India Sanitary Conference held at Bombay on 13 and 14 Nov. 1911, Calcutta: Superintendent Government Printing, 1912, pp. 133–37.
29. Cited in Martin Beattie, 'Colonial Space: Health and Modernity in Burra Bazar', *Traditional Dwellings and Settlements Review*, 14, 11, 2003, pp. 7–20.
30. West Bengal State Archives (hereafter WBSA), Mun. Progs, No. 165, Jul. 1883.
31. NAI, Home, Mun. Progs, No. 8–9, Jan. 1896. Calcutta Municipal Corporation, *Report on the Administration of the Calcutta Municipality for the Year 1889–90* (hereafter Calcutta Municipal Corporation, *Annual Report*). Lack of ventilation went together with diminished sunlight, making these houses 'veritable hot-beds of disease'.
32. 'Editorial', *The Statesman*, 21 Oct. 1890. Datta suggests that the concern with air may have been an individual predilection, with Simpson, the health officer from 1886 to 1897, being as obsessed with air as his predecessor Dr A. Payne had been with water. See Datta, *Planning the City*. I suggest, instead, a certain sequencing to this process, with most officials believing that provisioning of clean water took precedence over achieving purity of air.
33. Quoted in Datta, *Planning the City*, p. 185.
34. R. J. Minney, *Round about Calcutta*, London: OUP, 1922, pp. 20–21.
35. Datta, *Planning the City*, pp. 164–65. In his influential study of tuberculosis in India, Arthur Lankester similarly wrote of the continual failure of the Improvement Trust on the housing front: 'With all the improvements that recent years have seen, it still

remains all too true that the rooms of Indian dwellings, rich as well as the poor, are usually ill-ventilated, and they must in very many cases contain the germs of tuberculosis in an active form.' See Arthur Lankester, *Tuberculosis in India: Its Prevalence, Causation and Prevention*, Calcutta: Butterworth & Co., 1920, p. 131.

36. Cited in Sumanta Banerjee, *The Parlour and the Streets: Elite and Popular Culture in Nineteenth Century Calcutta*, Calcutta: Seagull Books, 1989, p. 60.
37. Cited in Partho Datta, 'Patrick Geddes and the Metropolis', Nehru Memorial Museum and Library, Occasional Paper, History and Society, New Series, 5, 2012.
38. Bose, *Tuberculosis*, p. 136. Not all concurred with this view. Mrs Kadambini Ganguli, among the first lady doctors of Calcutta, suggested that Indians were quite capable of reforms provided that adequate infrastructure—in this instance, more lying-in cottages for the poor—was made available, especially to the poor. Cited in 'Editorial', *The Statesman*, 16 Aug. 1890.
39. T. G. Hewlett, *A Paper on the Sanitary State of Bombay*, 1869, pp. 2–3. Hewlett served as the health officer and coroner of the city of Bombay at this time.
40. Tar was introduced to prevent the disintegration of roads when motor vehicles became widely used, while asphalt became common from the 1920s. Bombay Municipal Corporation, *Annual Administration Report*, 1939–40 (hereafter BMC, *Annual Report*), p. 198; 1942–43, p. 20.
41. Henry Conybeare, *Report on the Sanitary State and Sanitary Requirements of Bombay*, Bombay: Bombay Education Society Press, 1855, p. 5.
42. Cited in S. M. Edwardes, *The Rise of Bombay: A Retrospect*, Bombay: The Times of India Press, 1902, p. 300.
43. *Deaths in Bombay during 1863*, Bombay: Education Society Press, 1864, p. *viii*.
44. Edwardes, *Bombay*, pp. 323–24.
45. Ira Klein, 'Urban Development and Death: Bombay City, 1870–1914', *MAS*, 20, 4, 1986, pp. 725–54. One of the peculiarities of Bombay, Klein points out, was that much of the population was concentrated in the central wards of the city whose foul, airless slums presented an unhealthy contrast to the green, lush, beautiful tropical parks in some of its outer regions. Chandavarkar offers a fuller understanding of why this may have been so, including the

nature of the labour market that necessitated that workers live near the mills, the structure of the neighbourhood, with its systems of credit and protection, etc. See Rajnarayan Chandavarkar, *The Origins of Industrial Capitalism in India: Business Strategies and the Working Classes in Bombay, 1900–1940*, Cambridge: Cambridge University Press, 1994.

46. Cited in Kidambi, *Indian Metropolis*, p. 52.
47. BMC, *Annual Report,* 1895–96, pp. 582–83.
48. Cited in Kidambi, *Indian Metropolis*, p. 57. Jim Masselos writes that the plague thus constituted two incursions: one from the bacilli that attacked the individual and the other from the government. See Masselos, 'The Outside Inside: Incursions into the Marathi Household at the End of the Nineteenth Century', in *The City in Action: Bombay Struggles for Power*, Delhi: OUP, 2007, p. 126.
49. Kidambi, *Indian Metropolis*, p. 68.
50. Nikhil Rao, *House, But No Garden: Apartment Living in Bombay's Suburbs, 1898–1964*, Minneapolis: University of Minnesota Press, 2013, p. 25.
51. Cited in Kidambi, *Indian Metropolis*, p. 72.
52. BMC, *Annual Report*, 1902–03, p. 179. These 'light and air' discourses circulated in other parts of the empire too. See Cecilia Chu, 'Sanitation, Regulation and the Politics of Property in Colonial Hong Kong', in Robert Peckham and David M. Pomfret, eds, *Imperial Contagions: Medicine, Hygiene and Cultures of Planning in Asia*, Hong Kong: Hong Kong University Press, 2013, pp. 17–36; Jiat-Hwee Chang, '"Tropicalizing" Planning: Sanitation, Housing and Technologies of Improvement in Colonial Singapore, 1907–42', in *Imperial Contagions*, pp. 37–60.
53. J. P. Orr, 'Light and Air in Dwellings in Bombay', lecture delivered before the Bombay Sanitary Association on 27 Jun. 1912, Bombay: Times Press, 1912. The rule was to apply only to new construction.
54. Rao, *House, But No Garden*, p. 27. In the main though, the idea was to continue with working-class housing close to the old congested areas and, while doing so, looking for the cheapest land parcels.
55. City of Bombay Improvement Trust, *Administration Report for the Year Ending 31st March 1915* (hereafter BIT, *Annual Report*), Bombay: The Times Press, n.d., p. 96. On the different priorities of the Corporation and the Trust, see Kidambi, *Indian Metropolis*, Ch. 4; Rao, *Home, But No Garden*, Ch. 1.

56. *Report of the Textile Factories Labour Committee*, 1906 (hereafter *TFLC Report*), Bombay: Government Central Press, 1907, p. 3.
57. Cited in Priyanka Srivastava, 'Creating a Healthy and "Decent" Industrial Labour Force: Health, Sanitation and Welfare in Colonial Bombay, 1896–1945', Ph.D. diss., University of Cincinnati, 2012, p. 127.
58. Blair B. Kling, *Partner in Empire: Dwarkanath Tagore and the Age of Enterprise in Eastern India*, Berkeley: University of California Press, 1976.
59. What directed its use was its relative cheapness, about 12 to 13 shillings in the Calcutta market, as compared to the coal from Newcastle that sold for 25 shillings. D. H. Buchanan, *The Development of Capitalistic Enterprises in India*, New York: Macmillan, 1934, p. 548.
60. Cited in Kling, *Partners in Empire*, p. 64.
61. Samita Sen, *Women and Labour in Late Colonial India: The Bengal Jute Industry*, Cambridge: Cambridge University Press, 1999, pp. 14–15.
62. See Bengal (India), *Review of the Industrial Position and Prospects in Bengal in 1908 with Special Reference to the Industrial Survey of 1890*, Part II, Calcutta: Bengal Secretariat Book Depot, 1908, p. 3.
63. The citation is from a poem by Rupchand Bakshi, cited in 'The City Beckons', in Nilanjana Gupta, ed., *Strangely Beloved: Writings on Calcutta*, Delhi: Rupa Publications, 2014, p. 14.
64. Chandavarkar, *Origins*, p. 66.
65. Edwardes, *Bombay*, pp. 324–25.
66. D. E. Wacha, *Shells from the Sands of Bombay*, cited in Aditya Sarkar, 'Regulated Labour, Unruly Workers: The Making of Industrial Relations in Late Nineteenth Century Bombay', Ph.D. diss., SOAS, London, 2009, pp. 128–29.
67. Chandavarkar, *Origins*, p. 77.
68. Radha Kumar, 'City Lives: Workers Housing and Rent in Bombay, 1911–1947', *Economic and Political Weekly*, 22, 30, 1987, pp. PE-47–56. The first suburban service with steam traction was started in Bombay in April 1867, extended to Churchgate in 1870 and onto Colaba in the Fort area by 1873.
69. Sarkar, 'Regulated Labour', p. 159.
70. Sarkar, 'Regulated Labour', pp. 25, 84.
71. Sarkar, 'Regulated Labour', p. 90.

72. *Report of the Commissioners Appointed by the Governor of Bombay to Enquire into the Condition of the Operatives in the Bombay Factories and the Necessity or Otherwise for the Passing of a Factory Act* (hereafter *Enquiry Report*, 1875), Bombay: Government Central Press, 1875, IOR, V/26/670/85.
73. J. C. Kydd, *A History of Factory Legislation in India*, Calcutta: University of Calcutta, 1920, pp. 9–10.
74. *Enquiry Report*, 1875, pp. 57–59. At most, what the commission was willing to recommend was that some plan of ventilation approved by professional architects should be adopted by factories in the future. See *Enquiry Report*, 1875, p. 9.
75. Workers who were interviewed had no opinion to offer on the subject as none was asked, not even regarding cotton dust. The only questions posed to them were regarding trade-offs between shorter working hours and more wages.
76. A. G. Clow points out that almost all the local governments consulted, other than Bombay, had expressed their opposition to legislation being introduced regarding factory conditions. Eventually, however, the commission was of the view that it would be unfair if factories in one province were subject to restrictions that did not apply to other provinces. This was, naturally, much to the liking of the Government of Bombay. See A. G. Clow, 'Indian Factory Legislation: A Historical Survey', *Bulletins of Indian Industries & Labour*, 37, Dec. 1926.
77. On the 'natural' unity between the government and owners of capital in Bengal, see Dipesh Chakrabarty, *Rethinking Working Class History: Bengal, 1890–1940*, Princeton: Princeton University Press, 1989, pp. 74–85.
78. S. Gopal writes that the disappointment could be on either count: (i) that an Act had been legislated that, by itself, made Indian interests subservient to that of England or (ii) on account of the fact that it had not done enough to protect workers' well-being. Of course, the second argument was also seen by many in India as nothing more than a camouflage for the interests of English manufacturers. See S. Gopal, *British Policy in India, 1858–1905*, Cambridge: Cambridge University Press, 1965, p. 144.
79. Sarkar, 'Regulated Labour', p. 159.
80. Sarkar, 'Regulated Labour', p. 162.
81. *Report and Proceedings of the Commission Appointed to Consider the Working of Factories in the Bombay Presidency* (hereafter *Report and

Proceedings, 1885), Bombay: Government Central Press, 1885, IOR, V/26/670/86.

82. Not many considered this a problem, however, arguing that under dry Indian conditions, steam was absorbed and did not cause excessive humidification.
83. *Report and Proceedings*, 1885, p. 11; pp. 159–66. It was reported that in some such mills, it was impossible to see a person 20 feet away on account of dust.
84. *Report and Proceedings,* 1885, p. 205.
85. WBSA, Judl Progs, No. 89–96, Nov. 1892.
86. Kydd, *Factory Legislation*, p. 46.
87. See *Report of the Bombay Millowners' Association for the Year 1902* (hereafter BMOA, *Annual Report*), Bombay: Times Press, 1903.
88. Government of Bombay, *Provincial Report on the Working of the Indian Factories Act in the Bombay Presidency for the Year 1894* (hereafter Bombay, *Annual Factory Report*), p. 4.
89. Quoted in Shashi Bhushan Upadhyay, *Existence, Identity and Mobilization: The Cotton Millworkers of Bombay, 1890–1919*, Delhi: Manohar, 2004, p. 53.
90. 'Bombay's Slaves', *ToI*, 13 Sep. 1905. Emphasis mine.
91. The standard in England was 12 parts of CO_2 per 10,000 volumes of air.
92. *TFLC Report*, p. 18. Interestingly, the TFLC felt that dust was not much of an issue in textile factories, except in the case of a machine called 'willows' in cotton mills through which very dirty material passed.
93. On the move away from generalities about vitiated air to focus on specific, measurable and remediable risks in England, see Tim Carter, 'British Occupational Hygiene Practice 1720–1920', *Annals of Occupational Hygiene*, 48, 4, 2004, pp. 299–307, and 'Vitiated Air: A Victorian Villain?', *Journal of the Royal Society of Medicine*, 74, 12, 1981, pp. 914–19.
94. *TFLC Report*, p. 17.
95. India, *Report of the Indian Factory Labour Commission, 1908* (hereafter *Report of the IFLC*), Vol. II, Simla: Central Branch Press, 1908, p. 245.
96. *Report of the IFLC*, Vol. II, p. 232.
97. *Report of the IFLC*, Vol. II, p. 46.
98. *Report of the IFLC*, Vol. II, p. 43.
99. *Report of the IFLC*, Vol. II, p. 58.

100. *Report of the IFLC*, Vol. I, pp. 26–28.
101. One inspector even went so far as to suggest that improved ventilation would be a greater boon to the workers than the restriction of working hours. *Report of the IFLC*, Vol. II, p. 47.
102. *Report of the IFLC*, Vol. II, p. 115.
103. *Report of the IFLC*, Vol. II, p. 40.
104. *Report of the IFLC*, Vol. I, pp. 61–63.
105. Indeed, well into the 1930s, standards of cleanliness inside mills varied considerably, with fluff especially accumulating in those mills where cleanliness was done only on a fortnightly basis. See Cecile Matheson, *Indian Industry: Yesterday, To-day and To-morrow*, London: OUP, 1930, p. 32.
106. Bombay, *Annual Factory Report*, 1908.
107. Bombay, *Annual Factory Report*, 1909. The report also mentioned that there was no legal mechanism available to enforce the advice offered by inspectors.
108. Bombay, *Annual Factory Report*, 1910.
109. Bombay, *Annual Factory Report*, 1911.
110. BMOA, *Annual Report*, 1909.
111. Upadhyay, *Existence, Identity, Mobilization*, pp. 54–55.
112. See *Report of the Factory Ventilation Committee*, Maharashtra State Archives (henceforth MSA), File No. 1232 of 1912, General Department, Vol. LXII.
113. Thus bringing back a concern that had been flagged by the TFLC in 1906 but ruled out by the IFLC in 1908.
114. In 1921, T. Maloney was appointed as the advisor to the government on humidification, recommending standards based on kata-thermometric readings. See T. Maloney, *Humidification in Indian Cotton Mills*, cited in Royal Commission on Labour in India, *Evidence*, p. 58. However, even by the end of the decade, suggestions made by Maloney were being only gradually adopted. Specifically, on the question of dust inside card rooms, industry persisted in its view that given the unreliability of technology and high costs, not much was possible. However, it also found a more ingenious argument to wash its hands of the affair saying that while use of masks was an option, much would depend upon the voluntary efforts of workers as 'compulsion in such matters by employers might easily be misunderstood'. Government of Bombay, *Bombay Legislative Council Debates*, Vol. XXXV, 1935, pp. 1718–22; BMOA, *Annual Report*, 1939. On the other hand,

the concern with ventilation and humidification was also beginning to gain traction in other provinces thus allowing for comparison of humidity across styles of building, congestion of machinery and general factory conditions. See Delhi State Archives (DSA), File No. 15, B, Industries, CC Office, 1927. *Ventilation and Humidification in Cotton Mills in Delhi Province.*

115. Mrs Nathaniel Kindersley (Jun. 1768), cited in Datta, *Planning the City*, p. 138.
116. WBSA, Judl Progs, No. 1–4, Jun. 1882; No. 19–21, Feb. 1883.
117. WBSA, Judl Progs, Mar. 1880.
118. Thomas Oldham, the superintendent of the Geological Survey of India, opined that based on the proportion of fixed carbon they contained, the very best of Indian coals only matched the average of English coals. See Thomas Oldham, *The Coal Resources and Production of India*, Calcutta: Office of Superintendent, Government Print, 1867, p. 22–23.
119. Frederick Grover, *Report on the Abatement of Smoke Nuisance in Calcutta*; NAI, Home, Public B, No. 95, Nov. 1904. This would be as true of kitchens in rural areas as in the city, a paper on female occupations in Bengal by Babu Greesh Chunder Ghoshe, presented in 1868, noted, that even if the cooking room be large and spacious, it was 'impossible to disconnect it from the accompaniments of heat and smoke'. The consequence—'many delicate women have perished from disease contracted during the performance of this duty . . . many have permanently impaired their eyesight in the same cause!' See Babu Greesh Chunder Ghoshe, 'Female Occupations in Bengal', *Transactions of the Bengal Social Science Association*, Vol. II, Part I, Jan. 1868, p. 127.
120. In addition to the regular practices of cremation and burning of refuse that we discuss later, there could also be episodic events that led to great quantities of smoke being produced, for instance, in domestic *havan*s. These do not seem to find much mention though, either in government reports or in newspaper articles.
121. See David Arnold, 'Burning Issues: Cremation and Incineration in Modern India', *NTM*, 24, 2016, pp. 393–419.
122. The proposal had been first put forward in 1854 but not acted upon as the government feared that it would not be acceptable to Hindus.
123. Pande cites at length the opposition of one Baboo Ramgopaul Ghose who rhetorically posed the issue in terms of cultural difference:

'If the government can stop the practice on the banks of the Hooghly in Calcutta, because of a populous neighborhood, where will this stop? The cremation of the Hindoo may be prevented in the holy cities of Benares, Allahabad, and Huridwar … Call it custom or usage, or if you prefer, superstitious prejudice, I submit you are equally bound to respect it.' See Pande, *Medicine, Race and Liberalism*, p. 131.

124. William Eassie, *Cremation of the Dead: Its History and Bearings upon Public Health*, London: Smith, Elder & Co., 1875, p. 98; S. W. Goode, *Municipal Calcutta: Its Institutions in Their Origin and Growth*, Edinburgh: T. and A. Constable, 1916, p. 221.
125. WBSA, Judl Progs, No. 79–82, Mar. 1864.
126. Kaliprasanna Sinha, *The Observant Owl: Hootum's Vignettes of Nineteenth Century Calcutta* (*Hootum Pyanchar Naksha, 1862*), trans. Swarup Roy, Ranikhet: Permanent Black, 2008, p. 165.
127. See *Indra Nath Banerjee vs Queen Empress* concerning a burning ghat on the borders of the villages Mohanpur and Naihati in Bengal. The Calcutta High Court ruled that 'although a burning ground may not in itself be a nuisance … still a Magistrate will have jurisdiction if it is shown that such a *ghat* or ground is in such an offensive state, or that cremation is carried on upon it in such an offensive manner, as to be a source of injury, danger or annoyance to persons living in the vicinity.' Not only the religious sentiments of the people but also health and comfort mattered, and all cremation practices, it was ruled, must establish the necessary balance. See Indra Nath Banerjee vs Queen Empress (1898) ILR 25 Cal 425.
128. In his reminiscences of Calcutta, Cotton also makes mention of the cremation of the dead at Nimtollah, going on to add that 'the process has much to commend it from a sanitary point of view'. See Cotton, *Calcutta*, p. 344.
129. Goode, *Municipal Calcutta*, pp. 222–23.
130. Edwardes, *Bombay*, pp. 279–80.
131. Arnold, 'Burning Issues'.
132. Arnold also suggests that in the early decades of the twentieth century, the right to open cremation began to get mixed with nationalist claims about ritual practices and therefore it was only after 1947 that Western-style cremation began to gain some acceptance.
133. Incinerators had begun to gain widespread acceptance in England after the 1870s, though even here, smell and smoke from the chimneys

were subject to much local complaint. See Lee Jackson, *Dirty, Old London: The Victorian Fight Against Filth*, New Haven and London: Yale University Press, 2014, especially Ch. 1.

134. Conybeare, *Sanitary State of Bombay*, p. 35.
135. See L. W. Michael, *The History of the Municipal Corporation of the City of Bombay*, Bombay: Union Press, 1902, p. 262. Some would blame the smoke from incinerators as not inherent to the technology but on account of bad stoking, especially when the management of incinerators passed from private hands to the municipality. See 'The Cutchra Incinerator', *ToI*, 22 Sep. 1894.
136. 'The Calcutta Incinerators', *The Statesman*, 29 Jul. 1890. However, even as it was being endorsed, the fiscal efficacy of extending the Harrington incinerator to the whole city remained an open question.
137. 'Mr. Harrington's Incinerator', *The Statesman*, 31 Aug. 1890.
138. Arnold, 'Burning Issues'.
139. Goode, *Municipal Calcutta*, p. 163.
140. Arnold, 'Burning Issues'. However, faith in incinerators never entirely disappeared, being preferred over trenching for purposes of conservancy. See Government of India, *Indian Sanitary Policy, Being a Resolution Issued by the Governor General in Council, 23 May, 1914*, Calcutta: Superintendent Government Printing, 1914, p. 30.
141. T. Raleigh, *Lord Curzon in India*, London: Macmillan and Company, 1906, pp. 266–68.
142. Cited in S. Ehtisham Husain Rizvi, 'The Elements of Ghalib's Thoughts', trans. C. M. Naim and Muhammad Umar Memon, *Mahfil*, 5, 4, 1968/69, pp. 7–29, Asian Studies Center, Michigan State University. The first steamboat to navigate the Hooghly had been launched in 1823 about which the *Calcutta Gazette* reported that it seized the attention of local people with its 'smoke and speed'. Cited in Satpal Sangwan, 'Technology and Imperialism in the Indian Context: The Case of Steamboats, 1819–1839', in Teresa Meade and Mark Walker, eds, *Science, Medicine and Cultural Imperialism*, London: Macmillan, 1991, pp. 60–74.
143. Rudyard Kipling, 'A Real Live City', in *City of Dreadful Night*, New York: Alex Grosset & Co., 1899, pp. 5–7. First published in *The Pioneer*, 2 Mar. 1888. Such sensory experiences are important to take into account in developing our understanding of urban modernity. In his study of noise in modern Britain, James G. Mansell contrasts this sensed experience of an unnerving atmosphere, 'modernity

manifested in an audible form' as distinct from understandings of the modern as a set of ideas, institutions and practices. See James G. Mansell, *The Age of Noise in Britain: Hearing Modernity*, Urbana: University of Illinois Press, 2017.

144. Amiya Bagchi, 'Wealth and Work in Calcutta, 1860–1921', in Sukanta Chaudhuri, ed., *Calcutta: The Living City*, Vol. I, Delhi: OUP, 1990, pp. 212–23.

145. Dust, on the other hand, would be firmly associated with the Orient. In her account of water in Athens, for instance, Maria Kaika mentions the watering of streets in a bid to settle dust, which, it was hoped, would help dissociate the city from its recent (Ottoman) past. See Maria Kaika, *The City of Flows: Modernity, Nature and the City*, London & New York: Routledge, 2004, pp. 80, 94.

146. Bose was writing this in the context of his observations on bronchitis, pneumonia, etc., especially among those who lived in the vicinity of mills. Bose, *Tuberculosis*, p. 135.

147. Government of Bengal, *Proceedings of the Council of the Lieut.-Governor of Bengal*, Vol. XXXVII, 1905, p. 9.

148. Cited in Partha Chatterjee, 'Introduction', in Partha Chatterjee, ed., *Texts of Power: Emerging Disciplines in Colonial Bengal*, Minneapolis: University of Minnesota Press, 1995.

149. Raleigh, *Curzon*, p. 269. Writing of the American context, Stradling notes that while smoke was most encountered by those living in tenements in the industrial districts or near railroad stations and steamship docks, 'those who complained most bitterly often lived beyond the palls and looked upon the clouds from short distance … from a perspective outside the dense smoke pall, the gravity of the problem was clear'. Stradling, *Smokestacks and Progressives*, p. 23. In the Indian context, we may substitute middle-class Americans by the British administrative elite and bhadralok Bengalis who first took up the question of smoke as nuisance. There were, of course, contrary views too, for instance, the *Pioneer*, on 8 June 1903, writing: 'There appears to prevail in certain high quarters the view that Calcutta, being not merely a hive of commercial industry, but also the seat of Government and the capital of the Empire, must pay regard to its external aspect as well as to its internal prosperity. Whatever force may happen to be in this contention, the fact to be borne foremost in mind is that Calcutta is not a great commercial city because it is the capital of India, but, on the contrary, is the

capital of India because from the earliest British contact with the country it has been the centre of commercial enterprise … It is quite conceivable that jute factories, in their unobtrusive way, may contribute not less to the substantial happiness of the people at large than even a Victoria Memorial of white marble.'

150. 'Scientific Smoke', *The Bombay Times and Journal of Commerce*, 4 Mar. 1840.
151. 'An Intolerable Nuisance', *ToI*, 5 Oct. 1901.
152. 'Smoke Nuisances', *ToI*, 18 Dec. 1880. Though the names of the memorialists are not given, it is likely to have included some Indians too, the news of this being forwarded to the *Times of India* by one Vishnu Bhaskar.
153. 'Editorial', *ToI*, 31 Jan. 1882. Similarly, regarding Calcutta, the Smoke Nuisances Commission observed that this was an ever-present risk—'if you cook at an unsuitable "chula" smuts and soot settle on your food'—being both a health hazard and a nuisance. See Bengal Smoke Nuisances Commission (hereafter SNC), *Annual Report*, 1938.
154. 'Alleged Nuisance at the Nicol Mill', *The Bombay Gazette* (hereafter *BG*), 22 Mar. 1882.
155. 'Editorial', *ToI*, 31 Jan. 1882.
156. 'The Smoke Nuisance', *ToI*, 15 Oct., 7 Nov. 1901. The railways, of course, defended themselves vigorously, calling such complaints 'hypercritical'. See *The Pioneer*, 24 May 1896.
157. Douglas points out that women activists in Chicago deliberately targeted the smoke from the railways as this was nearer to ground and more easily felt by way of suffocation and fouling of clothes. See Douglas, *Cities*, p. 79.
158. Bengal SNC, *Annual Report*, 1920, 1926, 1927, 1928, 1937.
159. However, as we shall see through the course of this book, there were differential class and gender impacts that need to be borne in mind.
160. Bengal SNC, *Annual Report*, 1926. Total cost of smoke nuisance to Great Britain was estimated at roughly Rs 60 crore, though no such estimates are available for either Calcutta or Bombay.
161. Bengal SNC, *Annual Report*, 1930.
162. Bengal SNC, *Annual Report*, 1932.
163. Disciplinary proceedings in the form of fines and suspension usually resulted.
164. Grover, *Report on the Abatement of Smoke Nuisance in Calcutta*.

165. Anderson, 'Conquest of Smoke'; *Report of the Bombay Development Committee*, Bombay: Government Central Press, 1914, pp. *iv–v*. A proposal to move industries out of Calcutta was again made in the 1930s, with the caution that given wind directions, these be relocated on the eastern side. See Lt. Col. A. C. Chatterjee, 'The Smoke Demon of Calcutta', *The Amrita Bazar Patrika* (henceforth *ABP*), Health Supplement, 5 Feb. 1939.
166. Bengal SNC, *Annual Report*, 1926.
167. NAI, Home, Public, A, No. 221–22, Jun. 1904. This may be plausible, given that around the same time we learn of concerns in Delhi and Agra too about potential damage to Jama Masjid and Taj Mahal, respectively, on account of smoke from factories that were coming up in these cities. See 'Art and Smoke', *ToI*, 27 Dec. 1905.
168. NAI, Home, Public, A, No. 103, Nov. 1906.
169. Industrial smoke, in this reasoning, was about the failure to maximise value on account of the inefficient use of resources at the level of the individual firm. The cumulative efficiency of resource use, however, also had another dimension, though less pronounced, which was about the need to conserve the entire supply of coal that was deemed to be of 'vital national importance'.
170. However, the presence of public health specialists on Smoke Nuisances Commissions would be much resisted in the first two decades, as argued in Chapter 3.
171. NAI, Home, Public, A, No. 103, Nov. 1906. This conjoining of aesthetic and public health arguments would happen on more than one occasion, the Bengal SNC writing, for instance, of the need for developing a 'smoke conscience' in order that people could realise the indefensibility of tolerating prevailing atmospheric conditions from 'an aesthetic or hygienic viewpoint'. Bengal SNC, *Annual Report*, 1937.
172. NAI, Home, Public, A, No. 91–93, Feb. 1900.
173. Anderson, 'Conquest of Smoke', p. 297.
174. Even as we focus on the differences between regulatory practices in India and those in England and America, it is important to bear in mind that there were significant differences within Western countries too. Britain, towards the end of the nineteenth century, feared decline and associated the harms of smoke with decline in national and imperial fortunes. The USA, on the other hand, was a fast-growing industrial economy and sought to minimise the dangers of smoke with a view to producing healthy and beautiful

cities that befitted a society witness to large-scale progress. See David Stradling and Peter Thorsheim, 'The Smoke of Great Cities: British and American Efforts to Control Air Pollution, 1860–1914', *Environmental History*, 4, 1, 1999, pp. 6–31.

175. Smoke nuisance in England and the USA was the remit of municipal authorities. In India, by contrast, Smoke Nuisances Commissions were set up as early as 1905 in Calcutta, and then in 1912 in Bombay, replacing the municipal and police anti-smoke administrative mechanism envisaged in the anti-smoke laws of 1862/63. On regulating smoke in England through greater municipal oversight see Mosley, *Chimney of the World*, pp. 167–69; Thorsheim, *Inventing Pollution*, pp. 113–17. For municipal legislation in the USA, and their limits, see Stradling, *Smokestacks and Progressives*, Ch. 4.

176. On smoke and degeneration in England, see Bill Luckin, '"The Heart and Home of Horror": The Great London Fogs of the Late Nineteenth Century', *Social History*, 28, 1, 2003; Mosley, *The Chimney of the World*, pp. 102–07; for the USA, see Stradling, *Smokestacks and Progressives*, pp. 45, 51. Similarly in France, darkness on account of smoke had become an issue by the early nineteenth century, and one of the tasks before Baron Haussmann, who had been appointed prefect of Seine in 1853, was to make Paris less dark. See Alain Corbin, *The Foul and the Fragrant: French Social Imagination*, New York: Berg Publishers, 1986, pp. 133–34.

177. Angela Gugliotta refers to middle-class women environmental activists as municipal housekeepers who sought to extend the moral organisation of the home to the city as a whole. See Angela Gugliotta, 'Class, Gender and Coal Smoke: Gender Ideology and Environmental Justice in Pittsburgh, 1868–1914', *Environmental History*, 5, 2, 2000, pp. 165–93. For much the same reason, those in favour of smoke would often describe the efforts at regulating it as the making of 'woman's ordinance' that threatened jobs. See David Nye, 'Smoke Gets in Your Eyes: Pollution, Aesthetics and Social Class', *Reviews in American History*, 28, 3, 2000, pp. 422–27. Stradling notes four key features of the feminine definition of smoke: health, aesthetics, cleanliness and morality. The authority of these non-expert women's groups, however, seems to have declined somewhat after the First World War, with engineers gaining ascendancy and the question of economic loss gaining greater salience over that of domestic comfort. See Stradling, *Smokestacks*

and Progressives, especially Chs 3 and 6; Uekötter, *The Age of Smoke*, Ch. 3; Thorsheim, *Inventing Pollution*, p. 129.

178. In this, India seems to have mirrored more the situation in Germany, where too civic smoke abatement associations failed to emerge, leaving it to the bureaucracy to take the lead in abating smoke. See Uekötter, *Age of Smoke*, Ch. 2.

179. This was in contrast to matters related to river pollution about which we hear from various civic organisations, including the Calcutta Health Society and the Kashi Ganga Prasadini Sabha. The irony is that despite this, no law specific to river pollution was passed in colonial India. See Janine Wilhelm, *Environment and Pollution in Colonial India: Sewerage Technologies along the Sacred Ganges*, Abingdon, Oxon: Routledge, 2016.

180. Cited in Gyan Prakash, 'Body Politic in Colonial India', in Timothy Mitchell, ed., *Questions of Modernity*, Minneapolis: University of Minnesota Press, 2000, pp. 189–222.

2 | The Nuisance of Smoke, c.1860–c.1900

Matter in the wrong place is one of the foes that you have to deal with . . . black coal smoke, as emitted from the boiler chimneys of our factories, is included among those foes.[1]

Legislating for Smoke

In October 1862, the Legislative Council of Bombay passed an 'Act to abate the Nuisance arising from the Smoke of Furnaces in the Town and Suburbs of Bombay'.[2] The Act, in part, read:

> From and after the first day of January 1864, every Furnace employed or to be employed in the Town and Suburbs of Bombay in the working of Engines by Steam . . . shall in all cases to be constructed or altered so as to consume or burn the smoke arising from such furnace . . . after the 1st day of January 1864 [any person not doing this] shall . . . upon conviction . . . forfeit and pay a sum of not more than 100 Rupees nor less than 40 Rupees, and upon a second conviction . . . the sum of 200 Rupees.
>
> The words "consume or burn the smoke" shall not be held in all cases to mean "consume and burn all the smoke".
>
> Nothing in this Act shall be held to apply to Locomotive Engines to be used for the purpose of traffic in or upon any Railway within the limits of the Town and Suburbs of Bombay.[3]

The necessity for the Act, we are told, had arisen because the city had too many people residing in too crammed a space, who were being put to inconvenience by the smoke from the furnaces used in mills and manufactories that were increasing in numbers in the Fort and Colaba areas, as well as on the outskirts of the Native Town.[4] In addition, there was the smoke arising from private and public buildings where steam machinery was being used, which too added to the nuisance. Soon after, the Government of Bengal passed a similar Act for Calcutta and Howrah, for much the same reasons.[5] Starting 1 July 1864, all furnaces in the town and suburbs of Calcutta, the Act read, would have to be so constructed so as to consume or burn the smoke arising from them, with the usual caveat of escaping punishment if the best practicable means had been used. However, there were some minor differences too—steam boats plying on the river were granted the same exemption as the locomotive engines; the fine was reduced, being Rs 50 in the first instance and Rs 100 on subsequent conviction, and contrary to the municipal commissioner having been empowered in Bombay, in Calcutta, it would be the Boiler Commission that would take charge.[6] The Acts, in brief, were simple enough, allowing for prosecution on account of smoke nuisance if due care had not been taken to consume or burn the smoke arising from furnaces. This would not be easy, however, and not the least because the Acts provided a rather limited administrative machinery to implement it. In England, special sub-committees had been set up by the municipalities by the 1860s and 1870s to examine the smoke problem. In addition, large cities like Birmingham had granted powers to physicians and surgeons to bring in prosecutions for smoke offences. In Manchester, this was further consolidated by the deployment of police as extra eyes on the streets for observing smoke.[7] By contrast, in Calcutta and in Bombay, administrative arrangements continued to be rather limited right till the 1890s. The consequence was that observations of smoke remained rather uneven and prosecutions were no more than occasional affairs, often being long-drawn-out too. The 1862/63 Acts also made a distinction between 'consuming or burning the smoke' and 'consuming and burning all the smoke'. The operative clause would be 'as far as possible', with the Acts stating that

> the magistrate before whom any person shall be summoned may remit the penalties enacted by this Act if he shall be of

> the opinion that such person has so constructed or altered his furnace as to consume or burn, as far as possible, all the smoke arising from such furnace, and has carefully attended to the same.[8]

In the years ahead, much would hinge on how magistrates dealt with this clause. But this much remained constant: At the core of these regulations would be observations of dense black smoke for periods beyond an assumed 'normal', such normal to be arrived at by comparison with smoke emitted from the chimneys of other mills that were similarly located.[9] As for the design of furnaces and chimneys, some advice was on offer but more serious interventions would have to wait till after the setting up of the Smoke Nuisances Commissions in the first decades of the twentieth century.

Prosecuting Smoke

Calcutta

Public notice advertising the law was put out in Calcutta in June 1864, but for a full year nothing much seems to have happened, not because smoke nuisance had been prevented but because a suitable person to function as an inspector was yet to be found. Money was not available either, and under the circumstances, it was suggested that the inspector appointed for the purpose of boiler inspection also serve as inspector for smoke nuisance and the fines realised from the two activities be pooled together under a Steam Boilers' and Prime Movers' and Smoke Prevention Fund.[10] The approach of the Boiler Commission, in general, was more 'pragmatic than scientific', and much the same would be the approach towards abatement of smoke too.[11] The first of the cases to come before a magistrate concerned smoke from the chimney of the Eastman Company in 1865. The company did not contest the fact of smoke having escaped into the skies, but it defended itself by pointing to the inferiority of the Indian coal and its use of the best possible technology available that allowed for two-thirds of the smoke to be consumed. This was the maximum that could be achieved with available means, they contended, being supported in their assertion by engineers from other mills and from the Kidderpore and Howrah docks. There was not much by way of advice either by the authorities concerned, leading the magistrate to rule in favour of the company.[12] This was a pattern that was to repeat itself several times in the succeeding decades, that is, engineers of companies contested the observations made by the inspector of the commission leading to failure of prosecution, though for the moment what seems to have agitated authorities more was not so much this failure but the slow pace of action. Two more cases for offences under the Act were reported in 1866, the accused being acquitted in both.[13] In another couple of years, the police commissioner was still reporting that no fines had been collected yet.[14] Eventually, the first conviction was obtained in 1870, and among the early cases was that of one Rajmohun Bose in 1872 who successfully brought civil action against the workshop of the East India Railway Company, objecting to the smoke and smuts from it entering his house.[15] In the years ahead, special police was deputed to report on each chimney in the town, working together with the inspector of the commission to determine steps that could be taken to decrease the volume of smoke.[16] A notice was also circulated among mill owners stating that action would be taken against all offending parties and that mere mitigation of the nuisance, unless accompanied by permanent decrease in the amount of smoke, would be insufficient to meet the provisions of the Act. But not to much avail, it would seem. Instead, attention was drawn to the shortcomings of the law itself, the commissioner of police, who doubled as the president of the Boiler Commission, writing that smoke was purely a volume of steam, generated by coal in the furnace, coloured by invisible particles of carbon, incombustible and, therefore, inconsumable, rendering defective the clause 'consume or burn the smoke'. It was simply impossible to construct such a furnace, he wrote.[17]

The question of smoke seems to have attracted the attention of some other officials too, though again without much success. In 1877, the commissioner of Burdwan urged the government to extend the scope of the Act to districts outside Calcutta (Hooghly and 24-Pergunnahs), where large factories using steam power had begun to be established, a proposal that the government was not yet willing to endorse.[18] Some years later, the magistrate of Howrah expressed his seriousness about implementing the Act:

> I have paid personal visits to all the chimneys most seriously offending and emitting thick volumes of black smoke, and there are now prosecutions, ordered at my instance, pending in five cases. I have also again warned all other manufacturing firms by sending a copy of my preliminary proceedings with regard to the aforesaid five prosecutions. It is only by prosecuting that I can have the nuisance abated . . . I shall rigorously enforce the provisions of the law, and fine not only once for the commission of the nuisance, but again and again for non-abatement of the nuisance, till my orders are obeyed.[19]

Not all bought into these claims though, his superior officer noting that the magistrate's action had so far led to more excuses than convictions! The Government of Bengal too weighed in, writing to the magistrate about the continued emission of smoke by mills on the Howrah side of the river, with the chimneys of the Ganges Company being the worst offenders.[20] Quite obviously, the Act of 1863 had remained rather circumscribed, both geographically and practically. Indeed, just how little smoke figured in the calculations regarding sanitation even two decades after the Act had been legislated becomes evident in the recollections of Reginald Sterndale, vice-chairman of the municipality of the suburbs of Calcutta, some two decades later:

> There is another source of air pollution [that is, other than gases from traditional trades and ill-ventilation on account of overcrowded, windowless rooms], which, though as yet perhaps of little moment in this country, promises within a few years to add seriously to the contamination of the atmosphere of Calcutta and its extensive suburbs, as well as Howrah, Serampore and the towns along the banks of the Hooghly, and which it would be well that local authorities should not overlook, but endeavour to check in its infancy . . . This is what is denominated in Europe as the

> smoke nuisance, which pollutes the air by poisonous gases and by unconsumed particles of carbon, which not only blacken and discolour buildings and public monuments, but also choke and destroy vegetation, and exert a very detrimental influence on animal life. Every factory or place of business consuming a large quantity of fuel and creating smoke should be compelled to consume it.[21]

This was matter of special concern for Lieutenant Governor Sir Ashley Eden, leading him to appoint a special committee in 1879 to examine afresh the question of smoke nuisance in the city and what could be done about it.[22] After due deliberation, the committee attributed the enhancement of smoke in Calcutta to four factors: (i) an increase of about 10 per cent in the number of boilers attached to stationary engines; (ii) the substitution of coal, often of an inferior quality, for charcoal, wood and cow dung as fuel for domestic and manufacturing purposes; (iii) increase in the number of small steam vessels on the river and in the number of donkey engines for handling cargo; and (iv) increase in the number of sea-going steamers.[23] Eden was not quite inclined to consider the more minor sources, nor to legislate afresh to incorporate sea-going steamers into the Act, even though there was precedence for this in London and Liverpool. As far as Calcutta was concerned, he remained persuaded that the nuisance was mainly owing to smoke that 'poured forth continuously from morning to night from a small number of chimneys in the town of Calcutta and Howrah, especially the latter'.[24] And as regards this, the provisions of the existing law were sufficient, he believed, 'to preserve in the capital of India that purity of atmosphere which, until late years, it always enjoyed'.[25] All that was required was for members of the Boiler Commission to give special attention to the issue and to ensure that the procedure laid down by the law was carefully attended to in prosecutions that might be instituted.[26] Even so, ample caution would still be observed and the pace of alteration kept deliberate in order that no undue hardship was caused to industry. Immediate prosecution therefore was not what was being called for. Instead, what was recommended was a careful and systematic inspection of boilers, furnaces and chimneys in

Calcutta and Howrah under the aegis of the Boiler Commission, which would then offer appropriate advice as regards quality of fuel, method of firing and construction design.

The Boiler Commission made its renewed mandate clear at once. Stricter compliance with the law of 1863 would henceforth be insisted upon, but before instituting any prosecutions under that Act, the commission would afford opportunities for such remedial measures as the parties concerned may themselves be willing to undertake, including offer of professional advice if sought by the mill owners.[27] Several mills reportedly took on the offer, with the Howrah Mill, the Seebpore Mill and the Ganges Mill reporting that they were inclined to make suitable alterations based on such advice. Only one official seems to have remained dissatisfied—the magistrate of Howrah, who felt that the Boiler Commission was still not doing enough. In his understanding of things, the law of 1863 had given magistrates independent authority to initiate prosecution in the event that the Boiler Commission was unwilling or unable to do so. However, the new rules of 1879 seemed to have taken away this power, making it incumbent upon a magistrate to go through the commission, in his view entirely to the detriment of the cause of smoke abatement. References to the work being undertaken by the large mills, he was convinced, was no more than excuses for not prosecuting mills that had already received ample warnings in the past.[28] The Boiler Commission argued otherwise, reporting that more than 20 establishments, both on the Calcutta side and on the Howrah side of the river, had altered their furnaces and boilers in strict accordance with the suggestions of the commission's inspector. This was not all, for owners of all furnaces had expressed their willingness to do their best to abate the nuisance arising from smoke and to adopt any practical suggestions that the engineer may make. Under the circumstances, the commissioners had not thought it necessary to institute any prosecution till very recently, when Messrs Rally Brothers, Turner, Morrison and Company, and two other mills had been prosecuted.[29] The government too stood with the commission. It was for the inspector of the Boiler Commission to make the necessary observations, they noted, and only if he was satisfied about there being enough grounds for initiating action, would he approach the magistrate for obtaining necessary orders.

Given this administrative arrangement and the style of reasoning favoured by the Boiler Commission, the prospect of conviction naturally continued to remain rather dim.[30] The government made occasional mention of the fact that not enough was being done, requesting even stricter enforcement of the law.[31] But several practical difficulties remained, said the Boiler Commission. First, as had been mentioned before too, there was the wording of the law itself and its vagueness in determining the extent of the nuisance that was sufficient to justify a prosecution.[32] In England, the rule was to allow about 12 minutes of black smoke in the aggregate in each hour, while the Act in Calcutta merely stated that 'the words "consume or burn the smoke", would not be held to mean "consume or burn every particle of the smoke"', and the mere observation of smoke issuing from the chimney in large volumes for a few minutes did not quite suffice. The absence of suitable technology that could help consume the entire smoke only made matters worse. There was also the function of time and expertise. A furnace that the magistrate of Howrah or his police may have observed to be smoking excessively usually ceased to be a nuisance by the time the attention of the Boiler Commission's engineer was attracted to it.[33] And at any rate, the commission had to be convinced that they had the necessary evidence to suggest that the owner had been neglecting some precaution that he might have adopted, before ordering a prosecution. Once again, the magistrate of Howrah could not differ more. The practical difficulties being referred to were mere excuses and passing of the buck, he observed. Direct action by the observer nearest the mills was what was required, making mention in the process of the fines imposed by him on Messrs Burn and Company and on the owners of the Union Dock in May and November 1881, respectively:

> Messrs. Burn and Company ... had in October 1880 been warned by me, and repeatedly reminded both in writing and verbally. Their premises are close to my *cutcherry*, and

> the nuisance caused by their smoke for which they were prosecuted was palpable to my own senses. The legal evidence on which they were convicted was that of the police officers who were deputed by me to watch their chimneys. There was no real need to communicate with the Boiler Commission in this case, as I was perfectly acquainted with the whole circumstance. The defendants were of course convicted, as it was intended that they should be, or a prosecution would not have been ordered.[34]

Complaints against the Union Dock, he wrote, had been similarly dealt by him, and in a rare observation for an official, he added that while his actions may have been against government procedures, they were certainly more effective! There was only one way to go about things, he wrote, which was by way of keeping a constant watch on the chimneys and letting the owners and occupiers know that they were being so watched, and that they were therefore always liable to be informed against. The issue was also a sensitive one from a racial perspective, for the accused persons in Howrah had always been Europeans, making it imperative that the same magistrate who had ordered the prosecution also try the case himself.[35] Under current circumstances, he had endeavoured to deal with the cases with 'discrimination', but without leaving any doubt that the chances of acquittal were dismal, given that the nuisances being complained against were based on his own observation and that conviction was intended when the prosecution was ordered. This, he opined, certainly ran the risk of becoming indiscriminate. But, on the other hand, just writing to the Boiler Commission was 'sure shot guarantee of ineffectiveness'.[36]

The commissioner of Burdwan was persuaded, at least partially, regarding the necessity of empowering the magistrate of Howrah to initiate prosecution without necessary reference to the Boiler Commission.[37] There were two possible sources of black smoke, in his view, the first having to do with the construction of the boilers and the furnaces, and the second concerning the method of firing and the kind of fuel used. As far as the first was concerned, the matter was entirely technical and only the professional engineer of the commission could offer a definitive opinion on this. However, as regards fuel use and method of firing, it was entirely possible for the inspector of the commission to have inspected a particular set of boilers or furnaces and given a certificate of good condition, and yet the chimney continued to smoke. This, he believed, could be easily ascertained by the magistrate and the police, effectively and in time, and indeed, this is exactly how the law of 1863 had conceived the power of the magistrate with respect to smoke nuisance. The Boiler Commission continued to persist though that this would be a folly. The government was entirely at liberty to proceed along the lines suggested by the magistrate of Howrah and the commissioner of Burdwan, it wrote, but any such move would be detrimental to the administration of justice. Indeed, these officers hardly understood how the Boiler Commission functioned, for the certificate issued by them regarding a boiler and furnace was only as regards the risks of accident, not with respect to smoke. This they could do, that is, certify regarding the smoke-consuming arrangements at the same time that they certified regarding the boilers, but the real difficulty would still remain. And this difficulty had to do with the fact that there was no apparatus known that would satisfactorily consume and continue to consume the entire smoke of furnaces, especially in India. All that could be said at the moment was that there were various arrangements of furnace, fuel and careful stoking that could help reduce smoke. Hence, if any given chimney appeared to smoke unduly, it was by no means so easy, as was being assumed, to determine whether the smoke was due to defective construction of the boilers and furnaces, or on account of the method of firing or the fuel used. This would always require professional judgement. Determining whether the smoke emitted was excessive or not equally required professional judgement. It was not enough to say that a chimney smoked but also to establish that it smoked excessively, and it did so because the necessary efforts had not been made. This was a specialist's task, requiring a good knowledge of machines and techniques, not something that could be inferred on the basis of ocular evidence alone. It was for this reason that the government had entrusted the task to the Boiler Commission! The government, in turn, entirely endorsed the view of the commission:

> The object of requiring the Commissioners to be communicated with is to prevent mill and furnace owners being prosecuted on insufficient grounds, and the responsibility of instituting prosecutions laid on the Commissioners is this, that before authorising a prosecution they must satisfy themselves not only that a chimney smokes, but that it smokes to excess, and that the owners might have done more than they have done to abate the smoke. Sir Ashley Eden, after full consideration, decided that the Magistrate should not be allowed to act independently of the Boiler Commission in instituting prosecutions, and Mr. Rivers Thompson sees no reason to modify these orders.[38]

This then left one other aspect of the issue, viz., the area over which the Act should be applicable. The matter of extending the law to Hooghly, 24-Pergunnahs and similar districts had been first brought up in 1877 by the commissioner of Burdwan, who reiterated his views again in 1883, in a context where the Boiler Act had been amended to cover all places within 30 miles of Calcutta. Directly opposite the town of Chinshura in the 24-Pergunnahs side of the river, he noted, were the Goureepore Cotton Mills, the smoke from which, when the wind was from the east, blew into Hooghly and Chinshura. Below Calcutta, on both sides of the river, were several more, including Fort Gloster and the Bowreah Mills, all of which emitted dense volumes of black smoke constantly and did not seem to be under any control.[39] The comfort of the inhabitants of the suburban towns, he added, was as much a matter of concern as that of the people of Calcutta, and therefore it was imperative that something should be done in regard to the suburban mills too. The government was still not quite persuaded though. The time for such an extension had not yet arrived, they replied. Mills and factories outside Calcutta and Howrah, and their respective suburbs were not, as yet, so numerous, nor generally so situated, as to cause a nuisance in their vicinity by the smoke they emitted. Further, as had been pointed out on other occasions too, there was no apparatus known that would satisfactorily consume its own smoke. To extend the Act indiscriminately would thus risk mill owners being constantly subject to prosecution, based on the mere suspicion of a magistrate or of a police officer that the smoke arising from their furnaces was excessive. It was entirely possible that in time a better apparatus for the consumption of smoke than that currently in use may be discovered, just as in time, it might be that the number of mills in the country increased to an extent that rendered their being brought under control necessary. Necessary steps would be then undertaken, but in the meantime, 'looking at the difficulties by which the question is surrounded, the Lieutenant Governor prefers not to take action in the matter because occasionally the smoke from the Gouripore Mills, for example, is to some extent a nuisance to the inhabitants of Hooghly and Chinsurah'.[40]

The government, quite clearly, was disinclined to cause 'undue' hardship to industry. However, it did not entirely give up on smoke abatement, continuing to urge the Boiler Commission to adopt such measures as would protect the public from this 'preventable nuisance'.[41] Similar instructions were issued to the magistrate of Howrah, with a rather surprising mention that notices be issued under Section 290 of the Indian Penal Code (IPC) and not under Act II of 1863![42] Following this, the police made a local enquiry, submitting a list of offending mills in Howrah, noting that the degree of offence varied with character of the industry—the jute and cotton mills created greater smoke nuisance than foundries and iron works, probably because greater motive power was required and for a longer time. Due warning was issued and the magistrate, together with the inspector of the Boiler Commission, made visits to the various establishments, especially the Ganges Mill and the Seebpore Mill, which were the worst offenders. Once again, the view was that the excess smoke was on account of the absence of smoke-consuming apparatus, use of country coal and bad firing. Given the lack of trained labour, they reasoned, the focus ought to be on ameliorating the first two but not before the inspector of the Boiler Commission also noted that 'it is not for us to suggest any improvement; prosecution should be insisted upon, as the nuisance still continues and owners should adopt well-known methods'. The magistrate of Howrah was not as shy in offering advice, forwarding several suggestions to the agents of the two mills, and if there was no improvement even subsequent to this, prosecution would be initiated. In the event, action was taken against the Ganges Mill that had taken some remedial action, though belatedly, and therefore fined half of what was recommend in law, while the Seebpore Mill escaped being fined altogether, managing to convince that they had acted upon the advice offered.[43] In the years ahead, there would be even less to report, the actions taken being 'spasmodic rather than systematic'.[44]

Bombay

As in Calcutta, so in Bombay, initial activities in relation to smoke were rather slow to take off. Even more importantly, the municipal commissioner, and not the Boiler Commission, dealt with the question of smoke, and notices were issued under the Municipal Act and not under Act VIII of 1862.[45] The first notices were issued sometime around May–June 1867 to all those mills that were not using their furnaces to adequately suppress smoke.[46] The mills, in response, had their installations inspected by an engineer, in whose view a certain amount of smoke was bound to be generated even with the most skilful firing and, when once generated, could not be consumed entirely in the furnace. The point therefore could only be to reduce this amount to a minimum by using the approved style of furnaces and by careful firing.[47] The report, however, did not impress Arthur Crawford, the municipal commissioner, upon which the mills proposed that a committee of five engineers be constituted to undertake another inspection. This Crawford acceded to, hoping to thus demonstrate the regulation of smoke as a joint responsibility of the municipality and of mill owners who, as per law, were trying to adopt the best practicable means to reduce smoke. A new committee was duly appointed in December 1867, and they too reported along familiar lines—there was no practical plan that could abate smoke entirely, though with careful firing, ample boiler capacity and easy combustion this could be minimised. Some of the mills, the committee observed, were guilty in this regard, burning their fires beyond capacity, though they were willing to be corrected. This time, Crawford was suitably assured on the mills' intentions, and so the matter rested till 1873.[48] In brief, the first notice was served a full five years after the law was enacted, and a presumed intent on the part of mill owners to mend affairs was taken to be enough by the municipal authorities over the next five years or so.[49] The Act was clearly inoperative as yet allowing 'chimney-wallas', as the *Bombay Gazette* called them, 'to have their own way and kick up as much smoke as they pleased'.[50]

William George Pedder, Crawford's successor as municipal commissioner, however, had other intentions. He first asked the bench of justices for the assistance of an inspector. This was denied.[51] But Pedder was far from deterred, spurred perhaps by some strong sentiments expressed in the media:

> if the mill owners who fill the air with smoke by their parsimony and neglect have their rights and require protection, the general public who breathe a thick dusty atmosphere and are placed beyond redress have their rights also and require the protection of law. At present, the Municipal Commissioner is responsible for the smoke nuisance and we shall anxiously wait to see whether he will accept the responsibility which the law devolves upon him.[52]

Following this, Pedder wrote to the Alliance Mill in May 1873 that not only had there been frequent complaints by the public against smoke and soot issuing from their chimney but also that he had personally observed the same. Unless things improved, he would therefore have no option but to act against them. The warning, however, seems to have cut little ice, as evident in summons being eventually issued to the mill only in January 1875. The case, however, was an interesting one, not only as regards the degree of compliance with the law or otherwise but also because it brought to the fore the self-perceptions of both the prosecutors and the defence of mill owners, part technical and part social. Rather than being subject to penal action, mill owners asserted, their contributions to the city's progress must be duly underlined, with the Alliance Mill taking objection to the 'discourteous' and 'impudent' notice of the commissioner to the 'respectable' and 'influential' mill owners. The municipal commissioner did not deny them their prestige and high status either. Indeed, he responded, he was keen to promote their welfare and not paralyse the industry. But, he went on to add, 'the reason for my charging the mill owners is that I am bound to carry out the intentions of the Legislature, and also to prevent private individuals from resorting to litigation with respect to this nuisance'.[53] Mill owners and municipal officers could not be natural partners, in his view, refusing therefore to accede to a request by the Bombay Millowners' Association (BMOA) to appoint a committee of experts to look into the matter and settle it 'amicably'.[54] Interestingly, Pedder also read the law differently, taking greater note of the fact of emission of dense black smoke that annoyed neighbours than whether the mills had made the best possible efforts to curb it: 'It is not my business to enquire that the mill owners have used all appliances practicable. I believe that even if all practicable appliances were made use of, yet they could be charged with nuisance if their mill smoked.'[55] Notwithstanding this clear intent, however, the exercise continued to be fairly amateurish on the crucial issue of observing smoke, which could then be produced as evidence in court. Wassodew Ramchunder, an inspector from the Roads and Buildings Department of the municipality, described his experience thus:

> Mr. Walton [the executive engineer] told me to watch the mill. The correspondence between the agents of the Alliance Spinning and the Municipality was put in my hand with a remark that I should constantly watch the mill chimneys that were issuing smoke. I took my stand under a tree generally five or six hundred yards from the mill and watched the smoke. I was simply asked to observe smoke whether it was dense, dark and offensive. On the morning of the 5th of April [1875] ... Maneckjee Petit and Oriental Spinning Mills smoked as offensively as the Alliance. The Alliance smoked a minute more that the two mills....

> The Maneckjee Petit and the Bombay Spinning were the nearest mills from where I was sitting on the *chowkey*. The Oriental Mill was farthest from me.[56]

On another occasion, the municipal commissioner described his own inspections as follows:

> I saw the chimney of the defendant's mill smoking excessively. It was emitting a dense volume of black smoke. For about a quarter of an hour, or a little less, the chimney was smoking excessively the whole time. It was dense black smoke. I have watched chimneys for three years and half at intervals. In this case it was an exceedingly dense column of black smoke.[57]
>
> On the 27th of April last, on our round of inspection between half past seven and 8'o clock in the morning, I saw the Alliance chimney smoking excessively. I saw the chimney from Falkland Road. I saw a dense smoke issuing from the chimney. I then went to the mill on horseback. It took me about fifteen minutes to reach the mill. The chimney was smoking even then excessively. When I got to the mill I asked for the manager or the secretary. I was told that they were not in. I then asked for a clerk. A Hindu clerk came out, to whom I pointed out the smoke. He looked at the smoke. I remained at the mill for about 5 minutes. It was a dense black smoke, which I believe may have been avoided.[58]

The Alliance Mill, in turn, made much of these causal observations, arguing before the magistrate that they were surprised that the court was even willing to examine an instance where no time was taken, 'but simply casually guessed at the smoke being observed while those two gentlemen were out riding and occupied with other matters'. Inferences regarding density and volume of black smoke, they further added, were entirely subjective, lacking any means of testing accurately the density of smoke on this occasion.[59] The question of engineers as experts figured prominently too, as it would do in almost every other case concerning smoke, with the engineers of the government mint and those of the Sassoon Press Company, Fraser Miller and Company, etc., testifying that appropriate measures had been undertaken by Alliance Mill. Equally, the expertise on offer by the municipality was sought to be discredited, the Alliance Mill explicitly objecting to the municipal commissioner being examined as an expert as he was unlikely to know how smoke could be best reduced. The testimony of the executive engineer of the municipality was similarly objected to on the grounds that he was 'merely a civil engineer' and not a mechanical engineer who alone could be expected to have an opinion in a matter like this! By contrast, the engineer of their mills was 'a thorough engineer of large professional experience, who understood the ejection and consumption of smoke perfectly'.[60] The commissioner was far from persuaded though, countering the demand from industry for the appointment of a committee of experts, saying: 'I do not believe that whether the smoke can or cannot be reduced is a matter for experts only to decide. It is a matter of common sense and observation. I cannot see that it is a question for experts.'[61] And as regards Wassodew Ramchunder, the native inspector from the Roads and Buildings Department, racial bias got conjoined with the question of professional competency, the solicitors for the Millowners' Association advising them not to go into appeal on the ground that: 'It will be seen, on referring to the judgment, that the Magistrate has convicted the Company on the evidence of Mr. Pedder and Mr. Walton, and not on the evidence of the native who was employed by the Municipality to watch the mill.'[62]

Smoke inspection in the 1860s and 1870s was thus not about precision regarding the extent of pollution but more about assessing its possible legal determination. It was also not yet a matter to be dealt with by a specifically designated inspector either, the court making note of the contrast in this regard between institutional arrangements in English towns and those in India.[63] However, there was much that was shared too. As Whitehead notes with reference to the work of nuisance inspectors in Birmingham, the legalistic gaze of the nuisance inspector was not a form of governmental sight that assessed whether some absolute legal boundary had been transgressed but concerned instead an assessment of whether a pollution event could be converted into a legal prosecution.[64]

The manner of establishing the case would therefore not be by the extent of pollution but by the extent of it by comparison—if the Alliance Mill was producing more dense black smoke than the Morarjee Mill, then it implied that it was doing something that could be determined as being legally wrong. On their part, the companies sought advice from the municipality and from each other on how they could go about their task to ensure better outcomes, laid great stress on the fact that black smoke was bound to be produced at the time of firing, irrespective of the care taken, that strict municipal action in this regard would lead to thousands of workers losing their jobs and, in some cases, referring to the far greater volume of black smoke in the manufacturing towns of England. In the final instance, they too resorted to comparison but only to highlight assessments that favoured them—the Alliance Mill 'had the very best appliances for smoke consuming, that they always coked carefully, and that this mill smoked less than any other mill—the Morarjee Goculdas the only excepted'.[65] The magistrate was not entirely convinced. If the Alliance Mill smoked more than the Morarjee Mill, then something was obviously amiss, irrespective of the steps taken by the mills. If it smoked less, then there was not much of a case to be had. It had to be either the quality of appliances or the absence of adequate care in firing, and since engineers had testified in the affirmative to the former, the latter was the likely cause. It was on this count that the Alliance Mill should be prosecuted. As indeed it was.[66]

The dragging of the matter concerning the Alliance Mill and other mills from 1867, when Arthur Crawford had first issued summons to them, till almost a decade later, when they were fined in 1876, would suggest that periodic complaints and subsequent prosecution/remedial action notwithstanding, matters of smoke were not yet of sufficient urgency, even though they were of some interest. The mill owners feared that the fine imposed on Alliance Mill despite the long-drawn-out proceedings indicated that they were in the same position as before the commencement of the case. Their solicitors opined otherwise, suggesting that three important precedents had been set: (i) that the appliances which Alliance Mill adopted were, if not the best, at least as good as any appliance known for the prevention of the generation of smoke; (ii) that when a mill was firing, a certain quantity of smoke was bound to be generated, such that a chimney must emit thick black smoke for six or seven minutes at such times; and (iii) that if the evidence was confined to the observation that a chimney had smoked for six or seven consecutive minutes at intervals during the day, such evidence was not sufficient to secure a conviction under the Act.[67] Conviction, in other words, could only happen if the mill emitted dense black smoke for more than six to seven consecutive minutes after firing. This would then define for the moment the meaning of the phrase 'a mill must consume its own smoke so far as practicable'.

Following this, in May 1878, the municipal commissioner issued summons to the Sasson Spinning and Weaving Company which accepted the charge of emitting smoke but prayed for leniency lest an emerging industry be unduly restricted.[68] A few days later, Maneckjee Petit Manufacturing Company was fined Rs 50 on the basis of a complaint made by the governor who, having observed the smoke of the chimneys of the local mills, had instructed the municipal authorities to take action.[69] Such casual observation, the company countered, did not provide sufficient grounds for prosecution as no proof of harm to health of any resident had been offered, only to be rebuffed by the chief presidency magistrate who ruled that according to the letter of the Municipal Act of 1872 no such harm was required to be proven, only a nuisance sufficed.[70] In other instances, public protests, though limited, highlighted the harms to domestic amenities and to health. In 1881, a resident of Cumballa Hill urged fellow residents to jointly represent to the municipal commissioner regarding the smoke and smut nuisance and, in the event that he failed to act, to pool funds and approach the court directly.[71] Feelings were clearly aroused, as a couple of days later, another letter to the editor of the *Times of India* expressed the hope that this agitation would continue further and the municipal commissioner forced to act 'when the public feeling is expressed in a sufficiently forcible manner in the newspapers'.[72] A number of new mills were coming up, each new chimney adding its volume of smoke to the thickening air. The sea breeze might possibly disperse it during the day, residents argued, but early mornings could not but be a health hazard, the ride to the Kennedy Sea Face unpleasant, and the fog enveloping Malabar Hill entirely predictable. The *Times of India*, in turn, pressed on with its agenda for the appointment of a smoke inspector, as was the case in England, the desirable qualities of such an inspector in its view being the facility of visual discernment and the possession of moral virtue:

> [a person] who has the eye of an eagle and can tell among a forest of chimneys to whose mill any particular shaft belongs. He is no respecter of persons, and it is not an unusual thing to see a magistrate, or the mayor of a town, summoned for a smoke nuisance.[73]

On the other hand, there were the more cynical among Bombay residents, anticipating nothing but an eyewash:

> In England ... the inspector has been pronounced to be a failure. I have known a bottle of brandy close one eye of the inspector in that country, so how much would be required to close the two of your observant inspector on a very small monthly salary it is not difficult to conjecture, even allowing for the rigorous instructions and directions of some official of position.[74]

The plea, at any rate, went unheard, and observations continued to be made through ad hoc arrangements and for durations considered appropriate by the observer:

> I can see no reason to discredit the evidence by Mr. Kent. He appears to have received instructions from the Health Officer for the purpose of making observations of the chimneys; and he speaks to the fact that he went on 2nd December and watched five mills. He has also spoken to the fact that on the afternoon of that day he carefully watched the chimney of this mill for a period of thirteen minutes; and during that time the chimney emitted for eleven minutes, at intervals, dense black smoke; and that during two minutes two times there was only a little or no smoke. He again went there on the 8th December, at

> 2.55 o' clock in the afternoon, and made observations of the chimney for seventeen minutes. During those seventeen minutes for nine minutes at intervals the chimney emitted dense black smoke; and for eight minutes there was only a light smoke at times or almost nothing. Again, on another occasion, which was on the 20th January, he watched the mill at seven minutes before 8 o' clock in the morning. He watched it for one hour and two minutes, and says that during that time the chimney only smoked lightly; and that four times only it emitted dense black smoke, three times for a duration of half a minute, and once for a minute. I can come to no other conclusion from his evidence that he did observe a very considerable volume of thick black smoke coming out of the chimney during the time that he speaks of.[75]

And as cases continued to trickle into the court of the chief presidency magistrate, he persisted in arbitrating along similar lines as before—imposing a fine, though less than the minimum recommended by law, if credible witnesses reported seeing dense black smoke, while the proprietors and engineers of the mills, on their part, continued to insist on having made the utmost effort to minimise the smoke.[76] Interestingly too, cases began to be filed also with respect to smaller operations, either on account of bad furnace design and/or negligent firing, as provided in municipal law, or as resulting in a public nuisance, in which neighbours sometimes complained against one another. An instance of the first concerned the Prince of Wales Steam Flour Mill that was fined Rs 30 for negligent firing in 1883, the smaller fine being on account of the fact that this was a small factory, and this was its first offence at that.[77] A case of the second kind was reported in 1884 when one Munjee Cauurimbhoy accused his neighbour Sonebai of having erected three furnaces to be used as a factory for parching grain. Dense volumes of smoke issued from these furnaces, Cauurimbhoy argued, and the surrounding air was filled with particles of saw dust, husk and chaff, which also penetrated the interior of his house, causing a nuisance to himself and the 30-odd families who were his tenants.[78] The matter first went to the police court where a plea was made for the abatement of the nuisance. The magistrate, and afterwards the appeal court, however, declined to grant the order, arguing that the nuisance only affected individuals and not the public at large. Currimbhoy then moved the High Court, which agreed to arbitrate, ruling that Sonebai remove her furnaces and pay a sum of Rs 400 to Currimbhoy to cover costs.[79] Such instances of prosecution of smaller establishments, though rare at this time, would assume great importance in Bombay in the coming decades, especially after the setting up of the Smoke Nuisances Commission. Meanwhile, amendment in the municipal law would suffice to regulate, with new clauses added in 1888 to include smaller establishments such as bakeries, though railways continued to be excluded.[80] Following English precedence, the phrase 'so far as practicable' was defined to mean 'so far as it is possible consistently with the carrying on of the trade or manufacture in question', while nuisance was defined to include 'any act, omission, place or thing which causes or is likely to cause injury, danger, annoyance or offence to the sense of sight, smelling or hearing or which is, or may be, dangerous to life or injurious to health or property'.[81] But for much of the next decade or so, there seems to have been little by way of prosecution and the relationship between mills and the municipality remained one of cooperation, mixed with occasional warnings.

In 1891, the health officer wrote to the agents of Bombay United Company's Mill in Churney Road and the Khatau Mukoondji Mill at Byculla about the nuisance being caused by their mills, simultaneously bringing this to the notice of the Millowners' Association, asking whether, given the persistent complaints, any practical measures were being devised to abate this serious annoyance and danger to health.[82] The BMOA was careful to establish its willingness to address the issue, but it also made clear that the rules of the association did not permit any compulsory measures. An increased quantity of smoke was occasionally caused by the heavy firing, especially during the monsoon months, they maintained, though every effort was being made to induce members of the association to do all in their power to reduce the quantity of smoke emitted. In the meantime, the health officer was happy to observe that 'my warning to the agents seems to have had effect, as the nuisance has been abated'. In much the same vein, when complaints were made in 1896 regarding the continued issuing of dense black smoke from certain factories, the health officer was satisfied that 'the same having, in each case, been brought to the notice of agents and owners, the matter received prompt attention, and the nuisance complained was immediately abated'.[83] The necessity for taking steps to abate the smoke nuisance in Bombay was again brought to the notice of the BMOA in 1898, with a hint of prosecution under the Municipal Act.[84] And as before, the association assured that they had written to their members that smoke abatement was entirely in their own interest and hoped that the intimation would have the desired effect, making legal proceedings unnecessary. They also referred to the difficulties faced by industry, most importantly, regarding the high price of English coal owing to the prolonged strike in Wales, leaving them with no choice but to use inferior Indian coal, difficulties that were appreciated by the health officer.[85]

On other occasions, however, mill owners would be far less reluctant to own up to any significant responsibility, minimising the risks posed by their factories, observing that smoke nuisance was not only inseparable from manufacturing industries but also much less in comparison to smoke in manufacturing centres elsewhere. Indeed,

> from the fact that boilers and appliances of nearly all the mills at work in or near the city are of the most modern construction, and also that the coal consumed is, generally speaking, of a very superior quality, the smoke nuisance is by no means so great as it would be from a similar number of factories, say in Lancashire.[86]

Comparison was also made with smoke from government establishments and other users of steam engines, reference being made to smoke from the municipal pumping station (1891), the municipal incinerator (1894) and, even more frequently, the railways.[87] The relevant expertise in dealing with matters of smoke was also questioned.[88] Industry apart, residents also chimed in to bemoan the lack of effort on the part of the municipality: 'What a much less unattractive place this city would be if we had a Municipality which would rid its air of the quite superfluous amount of smoke emitted by its manufactories and quite superfluous amount of miasmatic fog emitted by its undrained marshes', wrote one resident.[89] The English press was not be too far behind either, the *Times of India* mocking the municipality for first suggesting that smoke was a perfect antidote for plague and then reversing its stand a couple of years later to suggest that the persistence of plague was on account of smoke in the city![90] And even when a major case was brought in, as against Sunderdas Spinning and Weaving Mills Company (March 1899), it was alleged that the municipality had become active once again only because 'some noise' had been raised in this regard by members of the Byculla Club!

The record of municipal officers, not the brightest to begin with, was clearly becoming open to deeper questioning by the 1890s. Even more significantly, the decade also saw a major change in the attitude of magistrates. We may recall that when cases of smoke nuisance had been first brought before the chief presidency magistrate, he had been willing to impose a fine if dense black smoke had been emitted on even one occasion, the quality of boilers, furnaces and other equipment notwithstanding. But as we approach the turn of the century, the magistrates' perspectives seem to have altered somewhat. They were still willing to go by the correctness of the observations made by municipal officials, but they were far less keen to punish, except when a particular mill was a persistent offender, as was the case of the Bombay United Spinning and Weaving Company:

> It is evident the chimney can be kept from smoking, and when the municipal officers asked that the forced draught might be put on, there was an immediate and marked improvement in the smoke. It is clear, therefore, that a proper use of the forced draught would tend to abate the nuisance. The use of the forced draught is to consume the coal perfectly, but the fiercer and clearer the flames the

> quicker is the coal consumed, and I can quite understand that in the present condition of the mill industry it is necessary to work as cheaply as possible. Therefore, there is a tendency to use common coal, and not too much of it. Yet if by doing so a nuisance is caused to the inhabitants of the locality, it is necessary to put a stop to it. I am satisfied that the managers of this mill have not used the means they have at their disposal for the prevention of smoke; they have been negligent, and for their negligence they must be fined Rs. 100, which is the maximum penalty under the Act.[91]

On the other hand, if satisfied that the mill, as a matter of routine, was diligent in the use of right technology and had made the necessary effort, an occasional lapse could be condoned:

> The municipality was bound to prosecute offenders who cause a nuisance to the public. There was in the present case [Soonderdas Spinning and Weaving Mills Company] no bias on the part of the officers, who did their duty in recording the facts and giving evidence regarding them. He [the magistrate] must conclude that on the 2nd March at the particular hour in the morning there was a nuisance. It was, however, not alleged that this was the constant and habitual condition of the mill ... The Act did not require that the mill should be a habitual offender, but it would not be advisable to prosecute for an exceptional offence which was almost accidental ... he trusted, however, that the mill would be more careful and try to avoid further accident of the kind.[92]

The Millowners' Association, not surprisingly, was quick to notice the changed attitude, appreciating the efforts of municipal authorities to promote the amenity of the city but ruing the fact that, on several occasions in the past, it had tended to ignore both the distinctively industrial character of the town and its open and airy location, with a vast expanse of sea to afford cool breeze.[93] Smoke that might matter elsewhere, in non-industrial towns or those inland, just did not pose the same harms in Bombay, and it was only the 'clamour of idealists' that had led to Bombay being treated as if it were 'primarily a fashionable watering place'! They were satisfied though that things were changing for the better, and 'a more practical view appears to have prevailed of late'.[94]

This change would be even more marked in Calcutta, and with it would come demands for a new law.

Reviewing the Law

The 'more practical view' of things, as we saw earlier, had also been true of Calcutta for some time. What was new in the 1890s was the opening up of the regulatory practice to greater scrutiny. The law, so far as offence and proof were concerned, appeared to be substantially the same in England and India. However, the procedure for prosecution was entirely different, with the enforcement of the law being entrusted to the Health Department and to inspectors appointed for the purpose in England but being with the Boiler Commission in Calcutta. The commission did have at its disposal inspectors who could be presumed to be in a position to observe when a chimney was a nuisance, but it was also becoming clear that evidence of 'similar nature' was not being entertained in India as in England.[95] The reference possibly was to a major judgement in the case of *Secretary and Engineer to the Boiler Commission versus Alexander Falconer, Manager, Central Jute Mills, Ghusery*, brought before the district magistrate of Howrah in 1892. The case, in brief, was as follows: On 27 April, G. E. Marklew, newly appointed as the smoke inspector by the Boiler Commission,[96] had noted smoke coming from the accused's engine chimney in question from 6 am to 7 am, from 8 am to 9 am, from 10 am to 11 am, from 2 pm to 3 pm, from 3 pm to 4 pm and from 4 pm to 5 pm. He prepared diagrams on the spot showing the colour of the smoke minute by minute during these six hours on that date, which, in turn, showed that during the hours in question there was dense smoke for 31 minutes in the first hour, 31 minutes in the second hour, 29 minutes in the third, 21 minutes in the fourth, 28 minutes in the fifth and 22 minutes in the sixth. The diagrams, we are told, were coloured and represented four different shades of blackness, with the darkest shade used to represent dense black smoke. There was little doubt, the magistrate held, that based on these diagrams, one would conclude that on the day in question the chimney of the accused's mill did emit excessive smoke, but there were several other issues that merited careful attention before arriving at a judgement regarding culpability.[97] Most importantly, there was the issue of 'blackness' itself, the manner of gathering this evidence, and what the law implied. It was difficult to give a satisfactory definition of what could be considered blackness, the magistrate held, and yet it was necessary that this be clearly understood for future guidance. There were degrees of density, and, in his view, the shades of colour offered by the Boiler Commission's diagrams gave a useful practical guide—anything darker than the second shade of black given in their diagram could be held to be dense. This, he believed, was both reasonable and not too severe on the mill owners. However, there would still remain some subjective dimension, the magistrate observing that in the final analysis, it was evident that whether the smoke could be described as dense black was a question of the senses. And in this instance, his own sense, and that of any unprejudiced person, he believed, clearly showed that the colour in the diagram was dense black.[98] The next issue that remained was how this evidence had been gathered. Marklew's submission was that he had observed the smoke from across the river Hooghly, at a distance of 1,125 yards, which was argued by the defence to be too distant for accurate observation. The magistrate overruled it, reasoning that dense smoke could indeed be seen from a great distance. Given then that the smoke was definitely dense black and concentrated in one place, a nuisance had been caused, and the question now was what could have caused it. The submission in this regard was that the nuisance was a consequence of bad stoking on the part of the accused's servants, an argument proposed both by the inspector and by the secretary of the Boiler Commission, though disputed by the defence. Two things found mention in passing—the fact that Bengal coal gave more smoke than English ones and that native stokers could not work like English ones—before being set aside on the grounds that English stokers were not much better either. Notwithstanding, there was enough experimental evidence to suggest that 12 minutes of black smoke was the the maximum that may be permitted as an outer limit.[99] This left just one more issue to be considered—whether the accused had made enough effort to combat the nuisance. The answer to this too, in this particular instance, was in the positive, the accused having appointed a special engineer to help reduce the smoke to a minimum. But had he done all that was possible? Here the magistrate was not quite satisfied, holding the accused responsible for negligent firing and therefore guilty of an offense under Act II of 1863.[100]

So matters rested for another 5–6 years, when the Boiler Commission decided to reappoint Marklew as a special inspector for smoke nuisance in June 1898.[101] The process pursued by the commission was described as follows:

> The Special Inspector is deputed under their orders to make observations of chimneys of the duration of one hour each to ascertain the density of the smoke emitted in that period. The results are carefully tabulated in diagrams, and a chart is kept for each mill showing the results of all the observations recorded. Where the amount for black smoke emitted is persistently excessive, the owners of the mill are served with one or more warning notices, and if these produce no improvement, prosecutions are instituted.[102]

Over the next six months, Marklew took 185 diagrams, prosecuted and fined five factories, two cases were reported to be pending in the Howrah magistrate's court and 52 mill owners were warned.[103] In another six months, 93 more diagrams were taken, 12 prosecutions instituted and 44 warnings sent. Improvements were reported in government establishments too, for instance, from 20 minutes of black smoke in an hour in February 1898 to 10 minutes in an hour in April 1899 in the Royal Mint, and from 15 minutes in an hour in November 1898 to 5 minutes by July 1899 in the port commissioners' hydraulic engine-house.[104] Together with this, the other tradition of offering constructive advice continued, the Boiler Commission forwarding pamphlets to mill owners with suggestions about how best to tackle smoke.

The principal causes of excessive smoke, they reiterated, were want of boiler power and lack of careful firing, and although the Indian coal was inferior to the Welsh coal, much might be done by introducing the system adopted on all private and state railways in India, viz., compensating the drivers and firemen for reducing the consumption of coal, which would also mean reduction in smoke.[105]

Notwithstanding their endorsement of the good work being done by the inspector and the perceived benefits that may accrue through sound advice, the commission, however, remained unsure of just how much could be achieved. The results obtained by the introduction of smoke-consuming appliances were negligible, they noted in December 1899—some had tried Meldrum's patent, in others, mechanical stokers have been introduced; some had purchased additional boilers and yet others appointed new foremen to supervise the stoking. But none of this had proved to be successful.[106] Indeed, in their view, there was no 'royal road' to the prevention of smoke, this depending on a combination of ample boiler power, properly proportioned flues and chimneys, good quality coal and good stoking.[107] Making matters worse was the increasing difficulty being faced as regards the evidence obtained and the possibility of this standing scrutiny in court. There was some success to begin with. In the instance of *Howrah Jute Mills* (1898) the magistrate upheld the validity of the Boiler Commission's diagrams as evidence, as also the fairness of the 12-minute limit proposed by it. This was followed by the case of *Empress versus J. Steen, Manager, Seebpore Jute Mill* in 1899. Once again, the 'facts' of the case were little disputed. On 12 January, under instructions from the Boiler Commission, Marklew had taken four diagrams between 3 pm and 5 pm, of which the first showed 23 minutes of dense smoke, the second, 17 minutes and the third, 11 minutes, as a consequence of heavy firing between 3 pm and 4 pm.[108] In Marklew's opinion, the mill had been generally fine (notwithstanding his own admission of three prior warnings), but in this instance, there were good grounds for prosecution. The magistrate also agreed that ordinarily the Seebpore Mill emitted comparatively little smoke, though on this occasion, there had been dense smoke between 3 pm and 4 pm. Indeed, the very fact that there was less smoke the next hour and on account of the generally good character of the mill, it could be reasoned that the law had been violated. He also saw no reason to dispute Marklew's evidence since, in his opinion, the inspector had considerable experience in watching smoke. Together, all this suggested beyond doubt that on the day in question 'the Mill sinned against its usual good character, and therefore there must have been negligence somewhere'.[109] A fine of Rs 10 was thus levied, the amount reduced on account of the fact that the occurrence had been exceptional. Marklew had been once again successful in getting a prosecution, but was this a pyrrhic victory? For notwithstanding the successful prosecution, an important dissonant note had been struck, regarding Marklew's experience and expertise, which came up short in the course of the cross-examination in the court:

> Gilbert Henderson, appearing for the Defence: Had you any experience of this work before the 12th January?
> Marklew: Yes. In 1892, when I was first appointed Inspector.
> Henderson: Previous to that you had no experience?
> Marklew: No. I was a mechanic in the Jamalpur Workshops for 12 years.
> Henderson: Did you then take observations of this kind?
> Marklew: No.
> Henderson: You have had no experience outside of Howrah?
> Marklew: No.
> Henderson: You have no experience in England?
> Marklew: No.
> ....
> Henderson: Have you ever worked as a Fireman at all?
> Marklew: No.
> Henderson: You are a mechanic?
> Marklew: I am a Mechanical Engineer.
> ....
> Henderson: Do you know anything of firing an engine?
> Marklew: I have never fired.

> Henderson: I may take it that you did not know when a boiler is properly fired and when it is not?
> Marklew: We have all got some idea of firing, but I have had no experience.
> Henderson: You cannot say that in this particular case there was negligent firing?
> Marklew: I cannot.
> Henderson: You cannot say whether the smoke was due to improperly constructed boiler or to negligent firing?
> Marklew: No, I was not there. I have not inspected the boilers.[110]

In time, this would serve as a damning indictment by Marklew of his own capacity to function as a smoke inspector. More immediately, he was in little position to establish whether the mills had done all they could in terms of boiler power, construction of furnaces, etc. So far, as we have seen earlier, the Boiler Commission had been keen to initiate prosecution through arguments of negligence, either in comparison to the mill's own record or in comparison to the functioning of other mills in the vicinity. In this line of reasoning, it had been enough to offer evidence of dense black smoke, as observed from a distance, with little or no effort made to enter a factory and examine the process internally. But by the turn of the century, this seemed to be finding little traction with magistrates. Thus, in *Emperor versus Burn & Co., Howrah*, no specific deficiency could be established—there was neither negligence nor was the duration of the smoke excessive—leading the magistrate H. LeMesurier to reason differently from his predecessors. According to the Act of 1863, LeMesurier observed, an offence could be on account of (i) using a furnace not constructed to consume its own smoke, (ii) negligently using such a furnace so as not to consume its smoke and (iii) not using *the best practicable means* for preventing smoke. And in his opinion, the prosecution was bound to prove that either the flues were insufficient or the boiler power too low, that the firing had been careless or that bad coal had been used. Great density and duration of smoke was presumptive evidence of this, and in extreme cases, the presumption could become practically conclusive. But when the defendants were found to use the precautions that the Boiler Commission had itself approved, the presumption was rebutted, and specific negligence or omission to use the best means must therefore be proved. This had not been done in this case, and so the prosecution must fail.[111] Similarly, in *Emperor versus S. E. Eustace, of the Calcutta Electric Supply Corporation* (May 1902), it was again held that dense black smoke, by itself, was not sufficient evidence, the prosecution being required to prove negligence or the failure to use the best practicable means for preventing or counteracting such smoke.[112] Quite clearly, doubts remained both about the purpose of the Act and the grounds of prosecution, with LeMesurier going on to add, 'I am not at all sure of the intention of this Act, and I shall be very glad if advantage is taken of this or any other case to obtain the decision of the High Court on the point of law.'[113] A year down the line, Fredrick Grover, smoke inspector of Leeds, seconded to India as an expert would conclusively observe: 'The two cases which appear to have terminated all attempts at prosecution under the present Act are those of *Emperor versus Burn & Co., Howrah* and *Emperor versus S. E. Eustace, of the Calcutta Electric Supply Corporation*.'[114]

Evidently, by the turn of the century, mill agents had been occasionally willing to accept that the smoke emitted by their factories was a nuisance but continued to press the point that often there was little that they could do about it. Indeed the burden of their argument was on the oppressiveness of the Act, which seemingly ignored that they were doing their best within the means available, taking utmost care in the construction of their furnaces and the management of the mills, held to be 'in the hands of certain of the leading and most reputable mercantile firms of India'.[115] They may not have been entirely successful in their efforts but this, in no way, invited the government, especially the local magistrates, to act 'arbitrarily' and in an 'illiberal' spirit towards them. By contrast, they drew attention to the provisions of the Public Health Act of 1875 and its presumed liberal approach and interpretation in England. Like the Boiler Commission, they too demanded amendments in law, not so much to clarify 'blackness' or the number of minutes for which such smoke ought to be permitted, but so that it was made more lenient. And so in 1899, the Jute Manufacturers' Association took its grievance to the Bengal Chamber of Commerce, asking it to approach the government. The long-term solution proposed was a new, more liberal Act, but till such time, as a preliminary measure, an expert commission was to be constituted 'for the purpose of examining into, and reporting upon, the question how far, and by what means, the emission of smoke from the Mills in the Town and Suburbs of Calcutta admits of abatement consistently with local conditions'.[116] Any such law that may be framed, they further argued, must apply as much to government mills as to private ones. There was also the issue of appropriate technology. Time and again, mill owners argued, they had approached the Boiler Commission to suggest alternatives for the prevention or reduction of smoke, only to be greeted by silence. On the other hand, there was 'continual harassment, anxiety and expense of these vexatious prosecutions'.[117] Finally, they were not convinced that Marklew was best suited for the job. He had, in his own testimony, admitted that he had practically no experience in making smoke observations, was ignorant of the methods suggested by the Boiler Commission regarding techniques of smoke observation, had no experience in the matter of firing boilers and was ill-equipped to determine whether the excessive smoke had been on account of negligent firing or faulty construction of boilers.[118]

Around the same time that industry was making a plea for a change in the law, the Government of Bengal wrote to Government of India regarding complaints of smoke nuisance received by them.[119] Ordinarily this should not have been so, for the Act of 1863 was an Act of the local government, the responsibility for whose amendment also rested with them. However, smoke was an exceptional issue, Government of India believed, as it concerned the amenities of Calcutta, which was a subject of direct interest to them.[120] Any possible amendments to the Act would therefore have to be with their assistance, for which they asked the local government to report on the appliances introduced to mitigate the nuisance and the number of mills that had introduced them. In response, Government of Bengal drew attention to the use of mechanical appliances and structural alterations, including special smoke-consuming apparatus and mechanical stokers, increase in the height of chimneys and enhancement of boiler power.[121] There was much to be satisfied about the effects of these, they wrote, the average black smoke from the 20 worst factories declining from 20 minutes in an hour in 1898 to 13.5 minutes the next year.[122] In the case of Burn and Company, the decline had been even sharper, from 35 minutes in June 1898 to 12 minutes on 15 March 1899. In the case of Howrah Mills, the decline was from 52 minutes in June 1898 to 19 minutes in March 1899, all testifying to the good work being done. But for the strict vigilance of the Boiler Commission, Government of Bengal proudly wrote, the mills would have relapsed to their old habit of reckless firing.[123] The smoke inspector had made 598 observations by 1900, they pointed out, 125 warnings had been served and 12 prosecutions instituted, all successful. There was also the practical dimension of the question, the available options being more palliative than remedial, though the government was committed to giving immediate trial to all technologies that may help suppress the nuisance.[124] Government of India begged to differ, noting that 12 prosecutions as against 116 warnings between 1898 and 1901 was hardly a record to be proud of. Warnings should have been followed by prosecutions, they argued, instead of issuing further warnings.[125] Indeed, the Government of Bengal's own account of how the Act had fared did not paint a good picture: after conviction, offending firms were allowed to go on just as before with simple admonitions; prosecutions were inadequately and negligently conducted; the necessary evidence required for conviction was not produced; the prosecutor did not give evidence of the last fine imposed; and Marklew was no expert.[126] It was the failure on the part of the local government to follow the policy of repeated prosecution of mill owners who disobeyed the law that accounted for the nonchalance of the mill owners. There was no question of relapse on the part of the mills, Government of India wrote, for there had never been a recovery in the first place![127]

It followed from these divergent assessments that the Government of Bengal was desirable that existing arrangements with regards to inspection, warning and prosecution should continue, as should the services of Marklew. However, they did concede that so long as the existing Act gave discretion to weak or lenient magistrates to discharge offenders if they pleaded that they had done their best, there would be difficulty in effecting much improvement.[128] Consequently, they were willing to contemplate an amendment to the Act so as to make the emission of black smoke beyond prescribed limits punishable, whatever the excuse offered. Such was the practice in England, and so should it be in India, with convictions secured without enquiring whether the stoker was skilled or not, nor what system of firing had been adopted, nor the quality of the coal used or the area and the height of the chimneys.[129] There was another, more delicate, issue though: the provisions of the Act of 1863 were identical to those of the Public Health Act of London of 1891, and it remained an open question whether powers ought to be assumed by government in Calcutta that were not bestowed on the executive in London.[130] The Government of India took a polar opposite view—the law was not strong enough and its administration a complete failure. Articulating this view most prominently was Viceroy Curzon himself, writing that the current inspection 'with its absurd diagrams, its spasmodic prosecutions, and its puerile fines' was a 'complete illusion'. Indeed, he would go so far as to suggest that there was a certain collusion between the local government and mill owners and the law was not enforced simply out of the fear that it would be unpopular with mill owners.[131] In keeping with this tenor, Government of India wrote to the local government in 1902 that the Act was 'of an antiquated type', containing no general rule-making powers and offending against 'all canons of sound legislation by the impossibility under its provisions of inflicting any really deterrent penalty except as the result of an almost interminable series of successful prosecutions'. There was an urgent necessity therefore for the Government of Bengal to adapt its legislation on the subject to 'modern requirements'.[132]

The question of law, in turn, was linked to another question: was the inspection defective, and if so in what respect? The matter was both personal and institutional. The Government of Bengal, through the members of the Boiler Commission, had expressed its faith in the 'industry, honesty, capacity and experience of their present Inspector', and therefore, not surprisingly, their desire to retain the services of Marklew.[133] Marklew had performed his duties with energy and efficiency, they wrote, within the limits imposed by the Boiler Commission. His personal qualities were also much to recommend for. The duties that appertained to his appointment were not such as to demand expert knowledge as much as good eyesight, keen observance, high energy, possessing some mechanical knowledge, having some experience of firing boiler furnaces and, above all, being honest.[134] In all these respects, Marklew was 'eminently qualified' and, in the opinion of the commissioners, it would be difficult to find another man better fitted to fulfil the duties of smoke inspector.[135] His diary, and the charts that he prepared, the former daily and the latter monthly, testified to his industry and to his zeal; his efforts to teach the firemen of the numerous boilers to stoke properly was immense, and 'if only they could be induced to work in his absence as they do while he is overlooking them, Calcutta would not have much to complain of'. The most they were willing to concede, therefore, was that his dual appointment as inspector of jute warehouses and as inspector of smoke nuisances was not the best possible arrangement and that he should be made a full-time smoke inspector. The Government of India was far from convinced, pointing out that the positive references of the commission were somewhat at variance with the evidence that Marklew had himself given in 1899 before the district magistrate of Howrah.[136] Neither was his record of prosecution entirely satisfactory. So even if he were to solely focus on his duty as a smoke inspector, as Government of Bengal was now proposing, there was little guarantee of any better results. Indeed the failure went deeper. The Boiler Commission, which supervised Marklew's work, had initiated criminal proceedings in a half-hearted manner. Their procedures were inexplicable—a firm after being successfully prosecuted had been allowed to repeat its offence to an almost indefinite extent with only formal warnings. And so it was doubtful whether the commissioners were themselves good judges as to what constituted an active and vigilant administration of the law for the prevention of smoke nuisance. Their view that smoke inspectors' duties did not demand expert knowledge of a very high order was also misleading. Sound technical knowledge, Government of India maintained, was an absolute necessity, for the nuisance could be on several counts—insufficient boiler room, improper construction of fireplaces in relation to boilers, inadequate arrangements as regards draught, improper construction of flues, improper stoking or bad management. Only a suitably trained and experienced inspector could ascertain how far, in each particular case, the nuisance was on the count of one or more of these.[137] The standards had also to be examined. In Glasgow, the standard was two minutes of black smoke, 3–5 minutes medium and 15 minutes of very light smoke, per hour. And if in India, the standard had to be brought down from the 12 minutes that now prevailed, the new proposal must be supported by expert opinion 'of sufficient weight to convince the magistrate and to oppose successfully the evidence of the engineers in charge of furnaces in Calcutta'.[138]

> We therefore adhere to our opinion that it is extremely desirable that the services of an expert in such matters should be procured from England ... we are anxious that the expert who may be sent out should be qualified by knowledge and experience to speak with real authority on the subject. We propose to postpone the question of amending the existing law until we have the benefit of his advice.[139]

With this, the die had been cast for framing a new law, based on the advice of the expert who could, after investigation, lay down 'with authority' the measures that should be taken to abate the nuisance while ensuring that the orders of the government were 'treated with respect by the mill owners'.[140] But the interests of the mill owners were not be ignored either, the government being quick to assure them that any law that would be framed would be done only after eliciting the opinions of the Chamber of Commerce, Jute Mills Association and other public bodies.[141]

'We do not want to arouse hostility on the part of the mill owners', wrote the secretary to the Home Department.[142] The expert was to assist both 'them' and 'us' to abate what was admittedly a nuisance by ascertaining the facts and making proposals for legislation with due regard for 'local conditions'.[143] The Government of Bengal was assured too that the expert would work under them, assisted by the inspector of factories. But the final report would be written in Shimla, in consultation with Government of India.

We meet this expert in the next chapter.

Conclusion

The making of the first laws of smoke in India, we have suggested, followed soon after similar initiatives in England. However, as things developed, major differences between the two contexts could also be discerned. Municipalities in England could count on a large and growing administrative staff. In India, by contrast, observations and inspection remained both rudimentary and amateurish. Again, while in England, law would offer clarity on the extent to which dense back smoke could be emitted, the law in India retained a certain vagueness through the first four decades of its implementation, restricting itself to consumption of smoke 'as far as possible'. Interestingly though, the differences that emerged were not only as regards England and the colony. There were also differences in the manner in which Calcutta and Bombay approached the question of smoke. The commissioner of police, in his capacity as the president of the Boiler Commission, was the chief authority in Calcutta. The municipal commissioner performed the same task in Bombay. And while in Calcutta, the Smoke Act of 1863 was evoked often and directly, in Bombay reference was more often to the municipal law, which contained wordings identical to the nuisance law. On the other hand, there was much that they shared in common too. In both cities, rhetoric mattered more than practical action on the ground, and there was an implicit assumption that at no cost should the abatement of smoke be at the expense of the mills.[144] Individual officers in both places—municipal commissioner Pedder in Bombay and the magistrate of Howrah in Bengal—would sometimes adopt a more proactive stance, only for subsequent action to be marked by greater caution. Thus, warnings remained more common than fines, and even this, when levied, was often much less than that mandated by law.

Industry, however, saw things otherwise, complaining that the law was acting harshly against them, especially given the poor quality of Indian coal, untrained firemen and the absence of suitable technology that could provide a satisfactory solution. Even more, they contested the authority of the inspectors, pitting them against the engineers hired by them, who were better trained and of European origin. By the turn of the century, magistrates too had begun to express doubts about the manner in which inferences were being drawn by the inspectors. Observing smoke was one thing. Whether this translated into a legal offence was another matter, depending on assessments of furnace design, nature of firing, etc. And these assessments of the technological ensemble, they felt, were beyond the capabilities of ill-trained inspectors. The Government of India, concerned about the amenities of Calcutta, got equally drawn into the subject, compelling Government of Bengal and the mill owners to agree to an amendment based on the enquiries of an expert, called out from England, whose recommendations would carry weight with all.

Notes

1. George Carruthers Thomson, engineer to the late Smoke Abatement Association, in a paper on the subject read before the Philosophical Society of Glasgow on 6 Feb. 1896. NAI, Home, Public, No. 91–93, Feb. 1900.
2. Legislative Council of Bombay, Act VIII of 1862.
3. Legislative Council of Bombay, Act VIII of 1862.
4. Government of Bombay, *Proceedings of the Council of the Governor of Bombay*, 1862, Vol. I, Bombay: Education Society Press, 1865, pp. 74–75.

5. In his account of smoke legislation in India, Anderson observes that it was almost without precedent or purpose. However, thought contextually, he adds, it was possible that the administrative and mercantile elites of Calcutta saw themselves in comparison with residents of London, where a serious campaign against smoke had begun to be mounted, that the smoke legislation was of a piece with other laws being passed by the colonial state that was increasingly prepared to intervene in everyday life and, perhaps, most importantly, because the sheer volume of smoke would have been on the rise after the coming of the railways, and the factories and steamships seen as creating a new urban condition. See M. R. Anderson, 'The Conquest of Smoke: Legislation and Pollution in Colonial India', in David Arnold and Ramchandra Guha, eds, *Nature, Culture and Imperialism: Essays on Environmental History of South Asia*, Delhi: OUP, 1995, pp. 293–335, especially pp. 307–08. As regards legislation in England, the Smoke Nuisance Abatement (Metropolis) Act had been passed to deal with matters in London in 1853, a decade or so before laws were being contemplated for Bombay and Calcutta.
6. Legislative Council of Bengal, Act II of 1863.
7. Mark Whitehead, *State, Science and the Skies: Governmentalities of the British Atmosphere*, Chichester, West Sussex: Wiley-Blackwell, 2009, pp. 45–46.
8. Writing about regulatory arrangements in Britain, Moran suggests that best practicable means, as distinct from an emission standard-setting exercise, became the signature tune for the flexible, empirical style of pollution control there too. See Michael Moran, *The British Regulatory State: High Modernism and Hyper-innovation*, Oxford: OUP, 2003, p. 45.
9. In other words, there was no defined standard against which 'excess' could be noted.
10. The Steam Boiler Act had been enacted in 1864, after a major accident in a factory the previous year had caused the loss of 13 lives. In England itself, the idea that regulation must be backed by a system of inspectors and inspections became established through the Factory Act of 1833. Regarding environmental pollution, such inspectorates became common following the Alkali Act of 1863.
11. In an interesting exchange regarding the resignation of Colonel Hyde, the Mint Master, from the commission for the inspection of

steam boilers, the Government of Bengal noted that his interest was primarily scientific ('and for which he desired that this committee be affiliated to the Manchester Steam Boiler inspection'), while the government was interested in 'only the prevention of accident by ordinary means of precaution'. See WBSA, Judl Progs, No. 140–42, 1865 and Judl Progs, No. 9–12, Jan. 1873.

12. 'Calcutta Police', *ToI*, 18 Sep. 1865.
13. WBSA, Judl Progs, No. 133–34, Dec. 1866. Government of Bengal, *Annual Report on the Administration of the Police for the Town and Suburbs of Calcutta for 1865–66* (hereafter Bengal Police, *Annual Report*).
14. WBSA, Judl Progs, No. 55–63, Jun. 1868.
15. Bengal Police, *Annual Report*, 1870; Anderson, 'Conquest of Smoke', p. 309.
16. WBSA, Judl Progs, No. 46–48, May 1877.
17. He also pointed out that the Act of parliament in England regarding smoke nuisance had become a dead letter for possibly the same reasons. The record of prosecutions in England around this time would seem to support such an assertion. The nuisance inspector of Manchester, for instance, reported over 4,000 observations and 140 summons issued in 1868 but only six cases taken before the magistrates and only five persons convicted. See Carols Flick, 'The Movement for Smoke-abatement in 19th-century Britain', *Technology and Culture*, 21, 1, 1980, pp. 29–50.
18. WBSA, Judl Progs, No. 41–43, Jul. 1877.
19. WBSA, Judl Progs, No. 67–72, Mar. 1879.
20. WBSA, Judl Progs, No. 31, Jan. 1880. The Ganges Company had been allowed three months within which to take necessary steps to abate the nuisance, but it seemed to have failed. In his response, the magistrate pointed out that while he had fined the company Rs 100, practical difficulties also needed to be appreciated. In particular, patented boilers that were in use did not work well and the company would have to be given more time to import new boilers from England.
21. Reginald Sterndale, *Municipal Work in India*, Calcutta: Thacker, Spink & Co., 1881, pp. 59–60.
22. WBSA, Judl Progs, No. 34–38, Jun. 1878 and No. 44–45, Jan. 1879.
23. WBSA, Judl Progs, Mar. 1880.

24. Anderson notes that this would not have brought much cheer though, as the chief culprit was not land-based engines but steam vessels. See Anderson, 'Conquest of Smoke'.
25. WBSA, Judl Progs, Mar. 1880.
26. Once again, the reasoning was that members of the Steam Boiler Commission were specially qualified to deal with technical questions, and they were also supported by subordinate staff whose special duty was to inspect all the steam boilers and prime movers in Calcutta, Howrah and the suburbs.
27. WBSA, Judl Progs, No. 19–20, Nov. 1880.
28. WBSA, Judl Progs, No. 45–48, Apr. 1880 and No. 6–9, Dec. 1880.
29. WBSA, Judl Progs, No. 1–4, Jun. 1882. In two of these cases, the defendants were warned and discharged, and in the other two, they were acquitted for want of sufficient evidence.
30. The annual police report for 1880 thus makes mention of only one case of complaint under the Smoke Nuisances Act in which one person was convicted, while that for the following year mentions two cases of which one was acquitted and another convicted. WBSA, Judl Progs, No. 22–25, May 1881; No. 34–37, Jun. 1882. Similar reports for 1878 and 1879 mention no complaints under this Act.
31. WBSA, Judl Progs, No. 19, Feb. 1882.
32. WBSA, Judl Progs, No. 1–4, Jun. 1882.
33. In other words, there was a certain elusiveness to excessive smoke in the absence of representational practices that could record the extent of it on a stable basis, making it a difficult element to regulate. This would continue to be so till Ringelmann's chart was finally adopted in 1905 (see Chapter 3).
34. WBSA, Judl Progs, No. 1–4, Jun. 1882. He did not, however, offer any reason why the fines imposed (Rs 5 in each case) were much less than that provided for in law.
35. Dipesh Chakrabarty points out that while mill managers and their assistants were often of working-class origins, their power and status was much amplified in the colonial context, as they became agents of rule in mofussil Bengal. This status was also reflected in the social exchanges that they enjoyed with the members of the bureaucracy. See Dipesh Chakrabarty, *Rethinking Working Class History: Bengal, 1890–1940*, Princeton: Princeton University Press, 1989, pp. 168–70. On the preference for European managers on

the basis of their presumed superior 'character', which reflected both a racial and class bias of mill owners, see Maria Misra, *Business, Race and Politics in British India, c. 1850–1960*, Oxford: Clarendon Press, 1999, p. 58. Over time though, more managers from other groups, especially Parsis, would find jobs as managers, their numbers equaling those of European mill managers in the cotton mills of Bombay by the mid-1920s. See S. M. Rutnagur, *Bombay Industries: The Cotton Mills*, Bombay: The Indian Textile Journal Limited, 1927, p. 294.

36. The engineer of the commission had just too many tasks to perform to be effective, he wrote, and the only way out was for a special assistant to be seconded to him to watch the smoke and note the periods for which black smoke amounting to a nuisance issued from a chimney, putting this information before the magistrate for appropriate action.
37. WBSA, Judl Progs, No. 19–21, Feb. 1883.
38. WBSA, Judl Progs, No. 19–21, Feb. 1883. But there was a flip side to the issue too. The same year, when two complaints of smoke nuisance were brought before the deputy magistrate of Sealdah, he dismissed it on the grounds that the necessary authority from a magistrate had not been obtained. Neither the police commissioner, who gave the order to prosecute, nor the deputy commissioner were recognised as vested with the powers of the magistrate. Upon this, the necessary authority of the magistrate of 24-Pergunnahs was obtained, only for the case to be dismissed a second time because permission had not been taken to examine the furnaces, as was required under the law. 'Bengal: Boiler Inspection in Calcutta', *BG*, 27 Oct. 1883.
39. WBSA, Judl Progs, No. 19–21, Feb. 1883.
40. WBSA, Judl Progs, No. 65–66, Mar. 1883. In the American context, Harold Platt makes the interesting observation that compared with river pollution laws, there was an absence of a regional approach in anti-smoke laws. In India, of course, no such law relating to river pollution was ever passed. However, it may be suggested that when debates on river pollution began to emerge by the 1890s, there was some appreciation of the regional dimension of the question in India too. Anti-smoke laws, by contrast, remained oriented almost entirely towards specific cities though the area under control expanded over time. See Harold Platt, *Shock Cities: The Environmental Transformation and Reform of Manchester and Chicago*, Chicago: University of Chicago Press, 2005, p. 41.

41. WBSA, Judl Progs, No. 22–24 and 25–30, Jan. 1892. The Lieutenant Governor wistfully recollected the actions taken after Eden's intervention and wished that so must it be now. Once again, the number of prosecutions increased, but only minimally, and the demand grew for the appointment of a special inspector.
42. Section 290 of the IPC provided for punishment for acts of public nuisance that had not been specified otherwise in the IPC. It would be interesting to speculate about the extent to which remedies under criminal law continued to be a resource for contesting actions even after a specific anti-smoke law had been legislated. Certainly in many other cases of nuisance, having to do with sound, vibrations, loud music, etc., varied sections of the IPC and the Criminal Procedure Code continued to be evoked.
43. See Empress vs George Alexander, Manager, Ganges Jute Mill, Howrah. WBSA, Judl Progs, No. 27, Mar. 1892.
44. WBSA, Judl Progs, No. 46, Apr. 1898. Meanwhile, several mills continued to complain that their requests for advice from the Boiler Commission were being ignored while the magistrates continued to fine them. See 'The Howrah Mills Co.', *The Pioneer*, 30 Jan. 1899.
45. Section 229 of the 1872 Municipal Act of Bombay read identical to the Act of 1862: 'From and after the date when this Act comes into operation, every furnace employed or to be employed in the City of Bombay ... shall in all cases be constructed or so altered as to consume or burn or reduce as far as may be practicable the smoke arising from such furnace.'
46. 'The Smoke Nuisance Case', *ToI*, 11 Aug. 1876.
47. BMOA, *Annual Report*, 1875 and 1875–76, Appendix F.
48. Around this time, we also learn that J. Mackinlay, chief engineer at the government dockyard, had visited several establishments in England at the request of the municipal commissioner to assess how best to abate the smoke nuisance. Based on what he observed, his suggestion was that large grate surfaces and slow combustion helped in economising on fuel and in producing less smoke. See 'Remedies for the Smoke Nuisance', *BG*, 26 Jun. 1868.
49. Some notices were also sent by individuals to the Justices of Peace, for instance, the one from F. A. Nazreth complaining of smoke nuisance in the Kumbharwada district. *BG*, 27 Jul. 1868.
50. *BG*, 4 Mar. 1873; 8 Oct. 1872. In order to put some teeth into the Act, they suggested that the inspector of steam boilers and prime

movers should be authorised to perform additional duties of the inspector of chimneys and furnaces in Bombay.

51. *BG*, 4 Mar. 1873.
52. *BG*, 4 Mar. 1873.
53. Individual complaints, however, continued to be made. One Rustomji Ardaseer Wadia, for instance, charged Morarjeedas Goculdas Weaving and Spinning Company with causing a nuisance to his estate on account of smoke, dust and noise, as a consequence of which his tenants left the estate causing him pecuniary loss. See 'The Smoke Nuisance Case', *ToI*, 23 Aug. 1875.
54. BMOA, *Annual Report*, 1875–76, Appendix F.
55. 'The Alliance Mill's Smoke Nuisance Case', *ToI*, 15 Jun. 1876.
56. 'The Alliance Mills Smoke Nuisance Case', *ToI*, 22 Jun. 1876.
57. BMOA, *Annual Report*, 1875–76, Appendix F.
58. 'The Alliance Mills Smoke Nuisance Case', *ToI*, 8 Jun. 1876.
59. 'The Smoke Nuisance Case: The Case Closed', *ToI*, 27 Jul. 1876. The nature of the observation continued to remain much the same in the coming years.
60. 'The Alliance Mills Smoke Nuisance Case', *ToI*, 27 Jul. 1876.
61. 'The Alliance Mills Smoke Nuisance Case', *ToI*, 8 Jun. 1876. The request for a committee of experts had come from the BMOA, and though such a committee does not seem to have been formed, three engineers were reported to have visited the mills and made recommendations for improvement, which were reportedly carried out. See also 'The Smoke Nuisance Case', *ToI*, 11 Aug. 1876.
62. BMOA, *Annual Report*, 1875–76, Appendix F.
63. 'The Alliance Mills Smoke Nuisance Case', *ToI*, 27 Jul. 1876.
64. Whitehead, *Skies*, p. 46.
65. The engineer of the Morarjee Mill believed that the use of Cardiff coal was possibly one reason why this mill smoked less than others.
66. 'The Smoke Nuisance Case: The Case Closed', *ToI*, 27 Jul. 1876. The fine imposed was Rs 50, on account of smoke having been caused by negligent firing on the morning of 27 Apr. 1875. 'The Smoke Nuisance Case', *ToI*, 10 Aug. 1876.
67. BMOA, *Annual Report*, 1875–76, Appendix F.
68. The magistrate yielded, taking a lenient view and fining them Rs 20.
69. 'Smoke Nuisance Case', *ToI*, 11 Jul. 1878; 'The Smoke Nuisance Case', *ToI*, 6 Jul. 1878. It is not inconceivable that several mill owners simply accepted the low fines as the cost of doing business, without bothering to change their ways of working.

70. In early court cases in the USA, similar objections were raised—even if there was smoke but if no harm to health or damage had been proved it was not actionable. The matter was finally resolved in 1900 in the case of *Moses versus United States* in which the court upheld that no such proof was required. Smoke could be presumed to be nuisance per se. Stradling writes that with this, and a subsequent ruling in Missouri, the authority of states to empower municipalities to regulate smoke was affirmed, and the broad powers wielded by cities to improve air quality maintained. See David Stradling, *Smokestacks and Progressives: Environmentalists, Engineers and Air Quality in America, 1881–1951*, Baltimore: Johns Hopkins University Press, 1999, pp. 66–67.
71. 'The Smoke Nuisance', *ToI*, 22 Dec. 1881.
72. 'The Smoke Nuisance', *ToI*, 27 Dec. 1881.
73. In Bombay, they wrote, 'the mill-owners are important persons with considerable influence in the municipal body, and they seem to have a chartered right of wasting their fuel in coal smoke and thus annoying the public. But in the manufacturing towns of the north of England no respect is paid to persons.' 'Editorial', *ToI*, 31 Jan. 1882.
74. 'The Smoke Nuisance', *ToI*, 6 Feb. 1882. In this, the writer may not have been too off the mark, with reformers in England around the same time often suggesting that municipal commissioners were hardly likely to prosecute those on whose goodwill they were practically dependent for their employment. See Peter Thorsheim, *Inventing Pollution: Coal, Smoke and Culture in Britain since 1800*, Athens, USA: Ohio University Press, 2006, pp. 118–19.
75. 'The Police Courts: The Smoke Nuisance Prosecutions', *ToI*, 13 Feb. 1883.
76. On more than one occasion, mills also resorted to an interesting line of defence by way of suggesting that the inspector(s) might have seen dense black smoke of another mill. Thus in February 1883, the Mahaluxmee Spinning and Weaving Company suggested that it was actually the chimney of the Bombay, Baroda and Central India Railway that may have been observed as being at fault and not their chimney. 'The Smoke Nuisance Prosecutions', *ToI*, 10 Feb. 1883.
77. 'The Smoke Nuisance Prosecutions', *ToI*, 15 Feb. 1883.
78. 'The Smoke Nuisance: Manjee Currimbhoy vs. Sonebai', *ToI*, 18 and 19 Mar. 1887.

79. Private nuisance law, as it had developed since medieval times, basically protected property rights and amenities in which the person seeking damages had to prove that a special harm had been done to him. In the case of municipal law, however, the emphasis was on harm to a larger public, and it is possible that the police magistrate refused to hear the case initially on this reasoning.
80. These were included in Article 391 of the revised municipal law of Bombay, which mandated that furnaces must consume their own smoke as far as practicable, with the exception of locomotive engines used for the purposes of traffic upon any railway or for the repair of streets.
81. See the commentary on Section 391, MSA, File No. 34314, 1888.
82. BMOA, *Annual Report*, 1891. Interestingly, the health officer also enquired from the association whether it had any information that it could share regarding smoke abatement in manufacturing cities of Britain. The association replied in the negative regarding new appliances, while maintaining that careful firing and ample boiler space were the only possible remedies.
83. BMOA, *Annual Report*, 1896.
84. BMOA, *Annual Report*, 1898.
85. Though he would go on to add that when it was known that Indian or Japanese coal gave off more smoke during combustion than English coal, it was the duty of all mill agents to see that better care was taken of, and more supervision exercised, over the process of firing.
86. Cited in N. Benjamin, 'Steam Boilers and Industrialization of Bombay, c.1850–1900: A Techno-economic History', *Proceedings of the Indian History Congress*, 69, 2008, pp. 612–24.
87. Benjamin, 'Steam Boilers'.
88. As we saw above, there were two kinds of evidence to be considered—one regarding the nature of the appliances used and the other regarding the quality of smoke. Evidence on the first was collected by the assistant executive engineer in charge of the workshop, the associate member of civil engineers and a member of the Institute of Mechanical Engineers, and little was disputed. That regarding the extent of smoke was done by the divisional health officer, which was considered appropriate by the municipal authorities but often disputed by the mills.
89. 'Smoke Preventing Devices', *ToI*, 22 Feb. 1899.
90. 'A Remedy for Black Smoke', *ToI*, 14 Feb. 1899.
91. BMOA, *Annual Report*, 1899.

92. BMOA, *Annual Report*, 1899.
93. This was a familiar line of argument in England and the USA too, with distinction being made between what was 'natural' and to be expected in an industrial town and what could be deemed 'pure' in a rural or suburban context. See Stephen Mosley, *The Chimney of the World: A History of Smoke Pollution in Victorian and Edwardian Manchester*, Abindon, Oxon: Routledge, 2008, pp. 134–35.
94. BMOA, *Annual Report*, 1900.
95. WBSA, Judl Progs, No. 27, Mar. 1892.
96. This seems to have been a brief, ad hoc arrangement with Marklew, who also served as the inspector of jute warehouses, appointed on a long-term basis as a smoke inspector only from Jun. 1898.
97. NAI, Home, Public, A, No. 390–94, Jul. 1902.
98. NAI, Home, Public, A, No. 390–94, Jul. 1902.
99. According to the secretary of the Boiler Commission, each time the furnaces were fired, a certain amount of black smoke was bound to be emitted, but it should not continue for more than three minutes. And as furnaces were fired about every quarter of an hour, this gave 12 minutes to an hour. The magistrate held this to be fair as mill owners got their reasonable rights and the public got reasonable protection.
100. NAI, Home, Public, A, No. 390–94, Jul. 1902.
101. The appointment was renewed on several occasions over the next few years. WBSA, Judl Progs, No. 77–104, Nov. 1899 and No. 33–37, Jan. 1900.
102. WBSA, Judl Progs, No. 33–37, Jan. 1900.
103. WBSA, Judl Progs, No. 77–104, Nov. 1899.
104. WBSA, Judl Progs, Dec. 1899.
105. WBSA, Judl Progs, No. 77–104, Nov. 1899. As we shall see later, instead of compensation, stokers were more often subject to fines and other measures of discipline.
106. WBSA, Judl Progs, No. 72–76, May 1902.
107. WBSA, Judl Progs, Dec. 1899.
108. NAI, Home, Public, A, No. 91–93, Feb. 1900.
109. Thus, the comparison in this instance was not with other mills but with the mill's own past performance.
110. NAI, Home, Public, A, No. 91–93, Feb. 1900.
111. In his review of the law as it stood prior to his enquiry, Grover wrote: 'The English Act of 1875 is unsatisfactory, because neither the density nor the time during which such smoke may issue from

the chimney are defined. The Calcutta Act and that upon which it is modelled are still more unsatisfactory, because in addition to the above noted defect the construction of the furnace is so emphasized that although it is the obvious intention of the Act to prevent the emission of smoke into the atmosphere, such emission has not been held to be sufficient proof of bad construction to warrant conviction.' Frederick Grover, *Report on the Abatement of Smoke Nuisance in Calcutta*. NAI, Home, Public B, No. 95, Nov. 1904.

112. NAI, Home, Public, A, No. 390–94, Jul. 1902.
113. NAI, Home, Public, A, No. 390–94, Jul. 1902.
114. Grover, *Report on the Abatement of Smoke Nuisance in Calcutta.*
115. NAI, Home, Public, A, No. 91–93, Feb. 1900.
116. NAI, Home, Public, A, No. 91–93, Feb. 1900; and No. 221–22, Jun. 1904.
117. NAI, Home, Public, A, No. 91–93, Feb. 1900. In the existing state of affairs, they warned, the jute industry of Calcutta would be severely crippled and few capitalists would be willing to invest their money in concerns 'so hampered and harassed by Government'.
118. NAI, Home, Public, A, No. 91–93, Feb. 1900.
119. These complaints could relate both to large factories and to smaller establishments, as for instance complaints of smoke from oil mills that were located in densely populated neighbourhoods. See, WBSA, Judl Progs, No. 77–104, Nov. 1899.
120. NAI, Home, Public, A, No. 41–44, Mar. 1903. The exceptional interest of Government of India in the amenities of Calcutta had been first established with regard to the Maidan. See Calcutta High Court, Hindusthan Mitra Club and Ors vs State of West Bengal on 23 Apr. 2002, (2002) 3 CALLT 515 HC, 2002 (4) CHN 1.
121. Alon Tal suggests that from a narrow local perspective, the 'solution to pollution' had always been dilution. Raising the height of chimneys was part of this strategy of dispersal of polluted air. See Alon Tal, *Pollution in a Promised Land: An Environmental History of Israel*, Berkeley, California: University of California Press, 2002, p. 255.
122. NAI, Home, Public, A, No. 91–93, Feb. 1900.
123. NAI, Home, Public, A, No. 12, Mar. 1902. In fact, at one point during the course of this exchange of views, the Boiler Commission went so far as to suggest that the improvement was so great that there was no need for prosecution at all. WBSA, Judl Progs, No. 24–25, Feb. 1901.

124. NAI, Home, Public, A, No. 390–94, Jul. 1902.
125. NAI, Home, Public, A, No. 390–94, Jul. 1902. Seebpore Mills, they pointed out, had been warned 12 times but prosecuted only once. Burn and Company were warned three times in 1898, five times in 1899, three times in 1900 and twice in 1901 but not prosecuted till 1902. Jessop and Company had been prosecuted twice in 1898 and had since received 10 warnings but again not prosecuted
126. NAI, Home, Public, A, No. 390–94, Jul. 1902.
127. NAI, Home, Public, A, No. 12, Mar. 1902.
128. NAI, Home, Public, A, No. 390–94, Jul. 1902.
129. WBSA, Judl Progs, No. 1–6, Jul. 1902. To reiterate, in England, the mere fact of dense black smoke was seen as being enough for prosecution. But under Indian law, the existence of such smoke was considered merely *evidence* from which inference could be drawn about whether the law had been violated or not.
130. NAI, Home, Public, A, No. 390–94, Jul. 1902.
131. NAI, Home, Public, A, No. 12, Mar. 1902.
132. NAI, Home, Public, A, No. 12, Mar. 1902. Indeed, while introducing the Bill in the Legislative Council, it was precisely the limitations of the law as it currently stood that were highlighted: 'Bengal Act II of 1863 … has proved completely ineffectual. The reason for this is that in order to procure a conviction under the Act, it is not sufficient to prove that smoke is emitted in such quantities as to prove a nuisance. Under the Act it is now held necessary to formulate a charge of nuisance under one of the following heads: (a) "improper construction", (b) "negligent use" and (c) "absence of the best means for preventing or counteracting of smoke". These conditions involve detailed references to the construction and manner of use of furnaces, and it is impossible to prove negligence or faulty construction in such a way as to obtain a conviction in a Magistrate's Court.' See Government of Bengal, *Proceedings of the Council of the Lieut.-Governor of Bengal*, Vol. XXXVI, 1904, p. 155.
133. NAI, Home, Public, A, No. 390–94, Jul. 1902.
134. We may recollect that similar qualities had been listed by the *Times of India* in Bombay as constituting essential qualities of an inspector.
135. NAI, Home, Public, A, No. 390–94, Jul. 1902; WBSA, Judl Progs, No. 72–76, May 1902.
136. NAI, Home, Public, A, No. 390–94, Jul. 1902.

137. NAI, Home, Public, A, No. 90–91, May 1903.
138. NAI, Home, Public, A, No. 390–94, Jul. 1902.
139. NAI, Home, Public, A, No. 390–94, Jul. 1902.
140. NAI, Home, Public, A, No. 390–94, Jul. 1902.
141. NAI, Home, Public, A, No. 41–44, Mar. 1903. A meeting was called to discuss the modalities of such cooperation, which included representatives of mills and of government. No Indian representative was called, making it evident that the matter, at this stage, was viewed primarily as concerning two different types of Englishmen in India—mercantile groups and government officials.
142. NAI, Home, Public, A, No. 41–44, Mar. 1903.
143. NAI, Home, Public, A, No. 41–44, Mar. 1903. At the annual meeting of the Indian Jute Mills Association held in February 1903, several mill owners were also of the view that industry should cooperate with the government, hoping to have a free and frank discussion with the proposed expert to estimate the extent of the problem and seek his advice regarding how best to mitigate it. See 'The Jute Industry', *The Pioneer*, 25 Feb. 1903.
144. Indeed, this would go to the very heart of the assessment regarding the use of best practicable means. Writing of mid–nineteenth century Manchester, Mosley suggests that to contemporaries this phrase did not imply either the best conceivable method or the best available method but rather the method that could be installed at a cost considered reasonable. See Mosley, *The Chimney of the World*, p. 139.

3 Expertise, Empire and Smoke Nuisances Commissions, 1900–16

Lord Curzon and the Empire of Expertise

On setting out on his journey as Viceroy of India in 1899, George Nathaniel Curzon addressed his old Etonian mates on the responsibilities that lay before him. Among these was to ensure that India achieved the 'full measure of her growth', by 'slow but sure degrees', even as Britain continued to rule over her empire in perpetuity.[1] Curzon believed this to be Britain's manifest destiny, but to ensure which the imperial mission would have to be radically recast. The investment of capital and employment of labour upon railways and canals, in factories, workshops, mills and mines and on tea, sugar and indigo plantations were seen by him as being essential to this recasting.[2] Equally, the affairs of the empire would have to be conducted with greater efficiency, Curzon having a keen sense of the fact that outcomes of the competition among the great powers in the twentieth century would hinge not on the superiority of force but on their relative efficiencies.[3] The burdens of the government were being added to everyday, he observed, by 'tendencies and forces outside of government', forces that were difficult to resist but, equally, 'not powerless to control'.[4] And so in this background, he set out to articulate a new conservative agenda for India's progress, one in which autocratic and benevolent British rule would seek to secure the interests of the people of India through development of Indian resources while maintaining social and political stability.

Colonial powers had always promoted a version of development, initially through trade and then through investments, to exploit colonial resources and peoples. However, by the late nineteenth century, the British had also begun to think of colonial policy in terms of its capacity to promote development along 'national' lines, the implied meanings of this terminology constantly mutating through the latter half of British rule.[5] Reformers who had earlier concerned themselves with localised health and sanitation issues, now showed greater urgency for what would come to be termed 'development' within colonial settings, the quest for progress and efficiency, as suggested by Curzon, being part and parcel of this. In all this, Curzon might have been charting a new path, displaying a greater willingness on the part of the government to intervene in the economy than had been the case with his predecessors. Numerous administrative reforms were also undertaken, with a view to ensuring that technical subjects were dealt by special departments staffed by specialists, and not by the generalist civilians who had hitherto manned the Indian bureaucracy. A Board of Scientific Advice was set up in 1902 and several other experts brought in as an immediate outcome of Curzon's emphasis—an inspector general of agriculture, a chief inspector of mines, a government architect, a government electrical advisor, a sanitary commissioner, etc.[6] And among these experts was Fredrick Grover, the smoke inspector of Leeds, seconded to India to propose new measures for effectively combating smoke in Calcutta. Anticipating his arrival, Curzon said:

> I may inform you that we have an expert from England, even now on the seas, coming out here to advise us as to how we may combat this insidious and growing danger. I hope, when he comes, all those who are concerned in the enterprises that result in such excellent financial dividends at the expense of so much fuliginous deposit will join hands with us in the attempt to curtail a mischief which, if unarrested ... will before long destroy one-half of the amenities of Calcutta, and will permanently injure its incomparable beauty and charm.[7]

The Making of the Bengal Smoke Nuisances Commission

Grover arrived in Calcutta on 24 February 1903, setting out to conduct his enquiry over the next four months. His con-clusions reinforced that which had already been much commented upon: Calcutta may yet be consuming just 300–400 tonnes of coal, in contrast to the 4,000 tonnes being consumed by Leeds, but this was enough to have made the resulting smoke into a nuisance, and it would be even more so in the future.[8] It was imperative therefore that the different sources of smoke, the use of alternative fuels or new technologies by which this could be abated and the institutional arrangements by which change could be effected be addressed in a comprehensive manner. Grover identified six different sources, including smoke from native huts, on account of traditional trades and the manufacture of coke, smoke from shipping and locomotives, and that from modern industry.[9] Steam boilers, he wrote, could be divided into two classes—those used on river launches plying on the Hooghly in the immediate vicinity of Calcutta and those used on ocean-going steamers, which when not on their way into port or out to sea were frequently moored on the Calcutta side of the Hooghly and were at times particularly objectionable.[10] Small river crafts also contributed to the problem, especially if two or three of them happened to follow each other, and each was either adding coal to its fires or raking them. But this was a problem that could be easily remedied, through a combination of good quality coal, careful stoking and good chimney draught. The use of coke, instead of coal, was also a possibility. Much the same combination of things could also be used for reducing smoke when the ocean-going vessels were moored near the banks of the river. On the other hand, it was but inevitable that ocean-going vessels, when preparing for their outward journey, would emit smoke in the process of raising steam, about which little could be done, at least as yet.[11] Brick and lime kilns, Grover further observed, were also difficult to regulate and were a problem both in the neighbourhoods where they were located and for the pilot service of the river Hooghly, as the smoke was emitted at a fairly low level. Given the unavailability of any technological solution for these, he suggested, the only way of dealing with this source of nuisance was their prohibition in places where they were likely to be regarded as a nuisance.[12] A third traditional source was the manufacture of coke along the river banks by natives employed on coal wharves, finding a solution to which would not be easy either. To the extent that this found a ready market amongst the native population for domestic consumption, it remained an open question whether an effort to regulate it through legislation would have the perverse effect of leading natives to burn the fuel in their own huts where they would be immune from interference and, by so doing, contribute even more to the nuisance than if the coke were made in open air near the river banks.[13] Then there were the furnaces used for the heating of plates and ingots, and for the reheating of metal in the process of shell manufacture and the like, which, Grover believed, could be improved with the introduction of heated air, supplied in the correct position and in correct quantity, to ensure the perfect combustion of smoke. In the main, however, it was not these 'minor' sources of nuisances that were the object of Grover's enquiries but the smoke from mills and factories that were affecting Calcutta's air.

The problem of industrial smoke as Grover outlined it was as follows: Calculations based on the use of fuel within a 5-mile radius of the Government House suggested that approximately 320 tonnes of coal was being burnt each day.[14] Of this, roughly 1 per cent fuel passed into the atmosphere as smoke, that is, about 3.2 tonnes of smoke was being emitted each day. The result was that 'the individual chimneys, although scattered, emit smoke for an intolerable length of time and that in consequence there are local areas within the five-mile radius which suffer much from factory smoke'.[15] Moreover, to the manufacturer in India, there was little inducement from a commercial point of view to address the problem. To be sure, there was a certain loss of efficiency on account of smoke formation in a boiler furnace, but this was unlikely to be more than 4 per cent. Given that the cost of fuel was but a minor part of the total cost of jute production and given the cost of supervision, it was unlikely that the jute manufacturer would save much by becoming more efficient or by ensuring less smoke.[16] Under the circumstances, the only means for suppressing smoke were likely to be either a sense of public duty in keeping the atmosphere clean or a legislation to regulate industry. The first could not be counted upon and in the instance of the latter, there was the failure of the Act of 1863 to be borne in mind. The only possible way forward therefore was to legislate anew.[17]

Grover's observations on these different sources of smoke and possible remedies for the same were subject to a range of responses, from an outright denial of the seriousness of the issue to its silences on other possible sources and, importantly too, fulsome endorsement from many. Articulating the first was the magistrate of 24-Pergunnahs, seizing upon Grover's observation regarding the relative purity of Calcutta's air as compared to industrial cities in England to suggest that the smoke nuisance in Calcutta was not of a kind that called for searching and drastic remedies.[18] The secretary, Sanitary Board of Bengal, and the chairman of the Calcutta Municipal Corporation went even further. It was possible, wrote the former, that when the wind blew from the north-west, the smoke produced in Howrah might become a nuisance, but it was only for very short periods that the wind did blow from that direction. Then, again, it had to be considered whether all the particles floating about in the atmosphere were entirely due to smoke or whether they were the result of the dirt and dust from the streets of a large and bustling city. Given this, the nuisance or injury to health caused by smoke in Calcutta was certainly not so serious as to necessitate new legislation.[19] The municipal commissioner similarly found little in the occasional misty mornings of Calcutta that could be considered prejudicial to the health of the people of the city, and indeed, it was his belief that it was 'practically impossible to entirely abate it in a town where 400 tons of coal are daily burnt, and to preserve an Arcadian atmosphere in a manufacturing town'.[20] The master of the mint was also not persuaded, arguing that more experimentation needed to be done and till then 'with our present want of knowledge of the efficiency of mechanical appliances in India or the prevention or abatement of smoke, it would seem unwise to attempt legislative measures until some more practical and efficient means are available ... for the abatement of smoke'. He worried too that unless liberal and workable rules were framed, those responsible for the control and management of boilers, etc., would be rendered 'liable to frequent and vexatious prosecution'.[21] The British Indian Association followed suit, apprehending that the authority to impose fines in instances of dense smoke, irrespective of what caused it, was likely to lead to 'abuse and oppression'.[22]

A second perspective focused more on the limits of Grover's recommendations and/or the likely impacts of the same. The British Indian Association, for instance, regretted that Grover had ignored several other sources such as steam rollers plying on Calcutta streets throughout the year, the Calcutta Port Commissioners' Railway, the Calcutta Municipal Railway and the Eastern Bengal State Railway 'plying night and day all along the western, northern, and eastern limits of Calcutta'; the East Indian Railway and Bengal–Nagpur Railway at Howrah; the engine sheds at Tallah; and the press, electric and other installations, public and private, in various parts of Calcutta and the suburbs which often proved to be a nuisance and danger.[23] The larger consumption of coke being advocated, in their view, was also problematic as this would necessarily be costly and would unduly hamper already struggling industries. And if coke had to be manufactured anywhere near Calcutta, the process, in itself, if carried out in open air, was likely to add considerably to the nuisance, compared to which the existing manufacture of coke on the banks of the Hooghly for private domestic purposes by coal wharf coolies was infinitesimally small.[24] The Bengal National Chamber of Commerce was not so sanguine either—the manufacture of coke along the river, in its view, was certainly a source of nuisance. But Grover's suggestion, if given effect to, would increase the demand of coke, which would further accelerate coke manufacture in Calcutta, thus exacerbating the problem.[25] These units, they further pointed out, were generally owned by poor and middle-class people, and provided the necessary means of sustenance for a large number of people who could be considered 'the children of the soil'. Similarly, if brick kilns were to be driven away from the open fields on the river bank where they were currently located, it was likely that they would penetrate into villages and hamlets, where the density of smoke would be even greater and the smoke nuisance would be even more keenly felt.[26]

But there would be support too, especially from the president of the Boiler Commission who concurred regarding the necessity of enacting a fresh legislation that would do away with the current trend of adversarial litigation in which negligence was difficult to prove. The existing Act, he wrote, did not necessitate the employment of an expert as a smoke inspector, but it did so for the purpose of giving evidence for a conviction, and therein lay its weakness as an equally good expert could be employed by a mill owner to show that his furnaces were constructed to consume their smoke within the meaning of the Act. 'It is this elaborate and unsatisfactory conflicting expert evidence', he reiterated, 'that squashes all the cases, and will continue doing so under similar legislation.'[27] He was also in agreement regarding the nuisance from the widespread manufacture of small quantities of coke by the natives—not only at the coal wharves but also all over the town and in Howrah—which must be stopped by confining them to spaces outside Calcutta. Similarly, regarding brick kilns that ought to be confined to an area outside Calcutta, approximately 4 miles from the town and suburban boundaries.[28] Several members of the Legislative Council also endorsed the measure being proposed, pointing to the nuisance caused, especially in the densely inhabited northern wards. Even more importantly, the Bengal government decided to yield in favour of a new legislation, much along the lines suggested by Grover. The contrary arguments, they opined, did not carry sufficient weight against the long-established principle, as articulated in section 268 of the Indian Penal Code, that 'a person is guilty of public nuisance who does any act … which causes any annoyance to the public' and which declared that 'a common nuisance is not excused on the ground that it causes some convenience or advantage'.[29] Government of India, which had recommended the expert in the first place, expectedly concurred. The only question was what form should the new legislation take.

Grover had proposed that wise legislation would be one that would secure improved amenities for the public without hampering industry or unduly interfering with commercial prosperity.[30] There were several related issues of concern here: whether black smoke in excessive quantities should be declared punishable, irrespective of what caused it; whether the power to declare culpability should rest with the smoke inspector or with another body of experts; what would constitute adequate evidence and the extent to which government ought to dictate technological choices. Grover's own view was that the new legislation should clearly define the furnaces to be included, as also clearly state the standard of comparison for smoke densities, the time for which the smoke density defined by the standard could be emitted and the period within which necessary changes must be made. Standards, he further reported, could be arrived at in two different ways—by measuring density by weight of soot or by appearance of smoke, and given the difficulties involved in the former, the latter was the more practical option.[31] The method of doing this was also specified, in the form of the use of Ringelmann's smoke gauge,[32] and the suggested time limit was of 10 minutes for dark grey/black smoke, progressively declining as the technology improved. The proposed Act, Grover further suggested, ought to also provide for smoke inspectors, who would be empowered to enter any factory or place containing a furnace during working hours, and use and test appliances for preventing emission of smoke. And contrary to what recent judgements had ruled with respect to the 1863 Act, the opinion of smoke inspectors as to the density and length of time of smoke emitted ought to be admissible as a relevant fact and presumed correct until shown to the contrary. This also implied, contrary to the view of the Boiler Commission, that the smoke inspector must be technically competent, well-versed in the use of smoke abatement appliances, a man of resource and one who could be relied upon to exercise his judgement, such that he would be entitled to the respect of those whom he was called upon to prosecute as also secure the confidence of the magistrate, whose judgement must necessarily depend upon his technical evidence.[33]

The responses to this part of Grover's report were also mixed, and none more so than with respect to the powers of the smoke inspector. Writing in partial support of Grover's opinion was the president of the Boiler Commission. Under existing legislation in India, factory owners were little inclined to control smoke, he observed. It was only the smoke inspector's diligence that helped improve the situation, and hence greater power to him was most apt. On the other hand, he did not expect this to imply a diminished role for the Boiler Commission. Only the commission possessed the necessary technical knowledge on boilers and furnaces, the president reiterated, and so their role must remain central to any new Act being contemplated. The commissioner of the Presidency Division argued for a committee with one or two official members, with the inspector to work under this committee.[34] The magistrate of 24-Pergunnahs argued for a Smoke Nuisances Commission, with members of mercantile firms represented on it, which could serve as an advisory body and to which people dissatisfied with the inspector's report could come. Only such a body, he argued, could offer 'a valuable protection to a mill manager honestly and conscientiously in search of the best practicable remedies for the nuisance caused by his chimney'. The Bengal National Chamber of Commerce made a similar argument, proposing that the Act must be administered by men familiar with these industries as also with official procedures. The Government of Bengal was adequately persuaded, making a provision for the creation of a Bengal Smoke Nuisances Commission in the Act being proposed, while providing for necessary powers to the smoke inspector regarding inspection and testing of appliances.[35] The matter, when it reached the Government of India, elicited contrary opinions though. On the one hand, it was felt that the creation of an advisory board would be useful in acting as a buffer between mill owners and the executive branch. On the other, it could be argued that an intermediary body would mean loss of efficiency. As Curzon put it, 'designed to prevent friction, it will do so in the most effective of all fashions, namely by preventing action which is the parent of all friction'. But on this, his opinion failed to carry weight with Risley, the secretary to the government, being persuaded in favour of a commission, not least because it had the backing of the Jute Association and the Chamber of Commerce.[36] This finally left the question of the admissibility of the inspector's report as evidence. Once again, divergent views were expressed. It would be against ordinarily accepted principles of evidence, argued some, to put the onus of proving the inspector wrong on the defendants. Risley, however, was of the view that it would be reasonable to assume that the opinion of a competent expert was correct.[37] The Legislative Department concurred. Questions as to the density of smoke, in so far as they were not questions of fact, must be determined by expert evidence, they opined, and the inspector should therefore be entitled, as every other expert was entitled, to give evidence on it. His opinion on the matter and his statement as to the length of time during which smoke was emitted would constitute prima facie evidence and would be presumed correct in the same way that the evidence of every other credible witness was presumed correct unless displaced.[38]

Divergent views were also expressed regarding the extent to which the local government could insist on a specific appliance being used. In the past, we may recollect, mill owners had sought exactly such an advice, suggesting that it was unfair to hold them responsible for excessive smoke in the absence of correct technical recommendations. Grover, however, was not quite persuaded, his enquiries suggesting that there was no device that would serve the purpose of effecting the consumption of smoke under all conceivable circumstances.[39] The president of the Boiler Commission had also been of the view that construction of furnaces, flues, chimneys or apparatus concerned owners most, and it would be dangerous for the authorities to take up such consulting work.[40] However, to the extent that there was a suggestion that furnaces must be suitably altered, there was a possible reading that the government would recommend appropriate technologies. The Lieutenant Governor of Bengal was quick to clarify—at the present stage of things, he wrote, it might be difficult to insist on specific appliances being adopted. What was required was the penalisation of excessive smoke as ascertained by tests to be prescribed, and if this were done effectively, the appliances required to avoid it would be automatically adopted.[41] The Government of India was equally keen to rule out the prospect, fearing that prescribing appliances would grant great powers to the experts and bind the government to a recommendation that may not have the guarantee of success.[42] All that the government ought to be concerned with, they too argued, therefore, were results, leaving it to the mill owners to best judge the means that would be most effectual.[43] Curzon, however, would remain unconvinced, suggesting that *eventually* this would have to be the course adopted. The time may not be appropriate for now, he wrote, and so the Lieutenant Governor and local government would not be foolish enough to force prematurely an untried apparatus or the individual fad of a smoke inspector on any factory. 'But sooner or later an apparatus will be discovered, entirely suited to the conditions, and then the power to prescribe may be very useful', he opined.[44]

In this background, a new Act to abate the smoke nuisance in Calcutta and its suburbs was proposed, much along the lines suggested by Grover. Furnace was defined to mean (i) any furnace or fireplace used for working engines by steam or (ii) for the purpose of carrying on any trade, manufacture or industry. Concomitantly, excluded from this definition were furnaces or fireplaces used for household or domestic purposes, for raising of steam on ocean-going steamers within a defined period prior to their leaving the port or to their moving from one port to another, and for burning the dead.[45] The erection of brick or lime kilns or the manufacture of coke, in ovens or with special appliances, could be prohibited within a specified area.[46] Assumptions regarding possible oppression and coercion were considered unfounded, the Act authorising the imposition of fines when smoke was emitted in greater density or for a longer time or at a lower altitude than that permitted by rules, whatever the cause and whatever the appliances used for diminishing the nuisance. It also empowered the government to direct that furnaces be so constructed as to not cause smoke nuisance, such direction also applying to furnaces that were already erected. And to put all this into practice, it provided for the appointment of a Smoke

Nuisances Commission[47] and of a smoke inspector, giving to the latter powers of inspection while keeping him subordinate to the commission.[48] Rules regarding density and time limits were also provided for, and the minimum height at which smoke could be emitted was fixed at 100 feet, though with a proviso that this would not apply to existing chimneys if, in the opinion of the commission, they were of sufficient height for the purposes of the Act or to furnaces in steam vessels.[49]

With this, the Calcutta and Howrah Smoke Nuisances Act, 1863, stood repealed. Smoke was now to be regulated in an entirely different manner, with magistrates to take cognisance of an offence upon a complaint made by, or with the written authority of, the chief inspector within a period of two months from the date of the commission of the offence, such offence being described in terms of the time and height limits imposed, irrespective of cause. However, this was not without retaining the original impulse of doing so without hurting industry: 'I trust that this Council will be satisfied that while government in making an attempt to grapple with a steadily increasing evil, it has been most careful to do nothing that could seriously hamper any important industry.'[50]

The Making of the Bombay Smoke Nuisances Commission

In Bombay, things moved a bit differently and a lot slowly than Calcutta. Proceedings in Calcutta were closely followed, but municipal action remained sporadic, and a new law was not yet on the horizon. What remained available though was suitably purple prose to describe the city as it stood under smoke filled skies:

> A northerly wind on Monday morning provided for the inhabitants of the Fort an atmosphere which resembled that of one of the busiest English manufacturing towns before the operations of the Smoke Act had taught the manufacturer that it is not only possible to burn factory smoke but that it pays to do so. Malabar Hill was quite

> invisible, and the sun rose in a red fire composed of the unburnt products of our local boiler furnaces.[51]
>
> Every year, when the cold weather comes round, we are called upon freshly to deplore the unnecessary evils which are allowed to burden life in Bombay. View the City early any morning and it is seen that the Fort and all the low-lying areas are wrapped in an impenetrable smoke fog. This is partially dispersed when the sun rises and the breeze springs up, but all the day, from any commanding point, dense clouds of the foulest smoke are found pouring from scores of factory chimneys ... The crystalline atmosphere of Bombay is being fouled to a disgraceful degree, and the city, which ought to be, from its position, one of the healthiest manufacturing centres in the world, is being reduced to a condition which would shame the Black Country of evil repute.[52]

Some mills continued to be prosecuted too, as before. In 1904, a notice was served to the Alliance Cotton Manufacturing Company Ltd, conviction secured and the company fined Rs 75, while the manager and the engineer of the mill were fined Rs 10 each.[53] In another case, some years later, Jacob Sassoon Mill was fined Rs 100 for having allowed dense black smoke to escape for more than 10 minutes.[54] More than the fine, however, the case was important for laying down an important new principle. As we saw before, hitherto there had been a critical difference in the laws of England and those of India, the former specifying exact time limits for smoke emission, while the latter only asked for all practicable measures being adopted to consume the smoke, allowing mill owners an easy escape route in the event of prosecution. In this instance, however, the magistrate decided to go the English way, ruling that even after making allowance for the cleaning of fires and the inferior burning quality of the Indian coal, no chimney in Bombay ought to be allowed to send forth black volumes of smoke for more than 10 minutes in an hour. The municipal commissioner and the health officer were elated, hoping to secure more convictions, even if the amendment in law along the lines of the English Public Health Act may yet take some time coming.[55] And some convictions were indeed secured, with the agents of the Madhowji Dharamsi Mill prosecuted and fined Rs 300—Rs 100 for each occasion when excessive black smoke was emitted in December 1907–January 1908.[56] In other instances, though, securing conviction continued to remain difficult. These failures haunted municipal authorities, none more so than in the case of Girgaum Paper Mills to whom notice was served in 1906 based on the excessive smoke observed by the municipal engineer himself, and prosecution undertaken with 'great care and skill'.[57] The mill disputed the facts and, as in the past, got another expert engineer to testify that the firing had been done properly and that there was no mill in the city where the firing was lighter. Objections were also raised that no prior notice had been served to the mill and, at any rate, there was no independent witness to corroborate the statements of the municipal engineer. The question also was 'what was practicable' and on this, the magistrate held, there could be differences of opinion, the accused thus being eventually acquitted.[58] The health officer was deeply disappointed, writing of the 'slow, cumbersome and ineffective process', which could only inflict a fine on proof of nuisance without any order to abate or prevent its recurrence.[59] And even as regards this proof, not much could be gained unless the law was amended and definitive time limits fixed.[60] A case was also made for the appointment of an inspector for smoke nuisance with good technical qualifications, the Health Department having neither the necessary engineering knowledge nor the time to attend to the issue. A year later, however, not much had transpired on any of these counts, leading the health officer to suggest once again that the existing provision in municipal law was so much in the interest of the furnace owner as to be almost unworkable.[61] Nothing happened even then, though the municipal commissioner remained hopeful that the 10-minute limit having being laid down, the executive would be bound to make even greater efforts to prevent the nuisance.[62]

Meanwhile, campaigns in the press continued, though oriented differently, focusing more on technological solutions than on the amendment of law. Excuses regarding use of Indian coal and untrained firemen, the *Times of India* wrote, were merely that, excuses! For there were several mills that did manage to keep the smoke within limits, offering as example the Maneckji Petit Mill, one of the largest in India, whose chimney did not afford the draught required by the furnaces and yet there was barely any colour in the gases escaping from it.[63] The key to this, the *Times of India* believed, was careful firing with the use of mechanical stokers.[64] The industry begged to differ, pointing out that there was no appliance yet invented that could help consume all the smoke generated by a boiler.[65] Indeed, it was foolhardy to expect that the nuisance was a preventable one. All that the *Times of India* was offering, in the view of one mill agent, was 'the aspirations of some inventive engineer or the blandishments of his commercial traveler, all to the mill manager's foolishness, and to the mill owner a snare', going on to add 'as a cobbler sticking to his last, you may take it from me that good hand firing is the only remedy for excessive black smoke, and anyone who urges the contrary is merely pointing out a means of wasting money'.[66] This was hardly their assertion, the *Times of India* replied. Instead, all they had suggested was that there were appliances available that even when fitted to a defective steam raising plant, so reduced the emission of smoke that it ceased to be a nuisance. More boiler power was also a possible remedy and 'as guardians of the public health', it was the duty of the municipality to enforce its adoption.[67] On other occasions though, it also took to cajoling industry leaders into accepting the proposals for smoke abatement, suggesting that it would add no further expense, either to their capital outlay or to running costs. It was in the nature of the competition among mills that none would be willing to take the first step, but the situation could still be mended if the corporation engaged the full-time services of a qualified inspector, a person with an engineering background and wide experience of different kinds of boilers, engines and furnaces, along with an adequate staff, who could report on the adequacy or otherwise of the draught required to completely consume the smoke.[68]

Mill owners were not exactly amused at the prospect of granting greater powers to the municipal corporation, as was being suggested by the health officer or the repeated urging of the *Times of India* for specialist inspectors with a view to more effective prosecution. Instead of seeking greater powers to prosecute mill owners, they argued, the corporation would do well to first set a good example by remedying its own defects, such as curbing excessive smoke from the pumping station at Worli or the air compressor at Colaba.[69] As for inspection, the chief want was advice rather than prosecution, mill owners pleading that while they were 'extremely anxious' to meet the wishes of the municipality, they did not know how to do it.[70] Things could be remedied, however, if the corporation secured the services of the right kind of expert from England, resorting to prosecution only when the advice of the expert had not been followed.[71] Eventually, as the prospect of an imminent change in law began to dawn on them, mill owners also took to pointing out that smoke nuisance had not yet been eradicated even in England or in the USA, despite all the advantages that those countries had. Under the circumstances, there was no reason to believe that rigorous enactment, without suggestion as to a palliative, would meet with any greater success in India. Older arguments about the incapacity of the government to deal with the matter in its own establishments were restated, once again to suggest that 'if Government with its enormous resources could not abate the nuisance in quarters under its official control, it was hardly equitable to expect a higher excellence from the Indian manufacturing classes with their limited experience in such matters'.[72] The emphasis, in the plea of the mill owners, was thus to focus on the 'art of the possible', not 'the urging of the idealists', of the kind that had been done in Chicago and elsewhere in the USA where the law dealing with smoke abatement was administered by a commission that was advised by a smoke inspector with due engineering qualifications, partnering industry in eliminating smoke rather than prosecuting them after the fact of excessive emission.[73] Practically, this would mean that the government should first appoint a commission of enquiry, which could look afresh at the question of smoke nuisance in Indian conditions and recommend what, if any, amendments in law were required.

In this background, a Bill for the amendment of the Bombay Municipal Act was presented in the Bombay Council in 1910, with the corporation seeking greater powers to deal with the smoke nuisance. The issue itself was not in doubt: 'Smoke, dense or black, emitting from a furnace is admitted on all sides to be a nuisance.'[74] The only question was how to control this nuisance, the most obvious solution being the imposition of a maximum time limit: 'No person shall use or permit to be used any such furnace as that it shall send forth from any chimney black smoke for more than eight minutes in the aggregate in any one house.'[75] Fazulbhoy Curruimbhoy Ebrahim, speaking as a representative of industry, countered that this was too hasty and too stringent, suggesting as before that the government must first appoint a suitable committee and await its recommendations before initiating change. And if a limit was to be at all imposed, it should be on the liberal side—15 minutes per hour rather than 8 minutes being proposed, bringing it closer to 12 minutes permitted in England where better coal was available.[76] However, before the question could be discussed any further, the matter took a completely different turn. The municipality, which had hitherto sought greater powers, was now entirely ruled out of the equation, and a proposal was made for the setting up of a Smoke Nuisances Commission like in Calcutta, based on the recommendation of the Select Committee to which the Bill had been sent, and which was assisted in its task by three representatives of the Millowners' Association:

> The Committee . . . are of the opinion that the enforcement of a clause like the one in the original Bill introducing a time limit will be attended with practical difficulties that will make its utility for the purpose of minimising the smoke nuisance questionable and they have therefore recommended that instead of dealing with the subject in this incidental manner it would be better to make the remedying of this evil the subject of separate legislation on the lines of the Bengal Smoke Nuisances Act of 1905.[77]

Sir Pherozeshah Mehta begged to differ, and differ strongly. The corporation's hands, he argued, had been forced by the weakness of existing legislation. The clause regarding the adoption of 'best practical means' had provided escape routes for industrialists when they caused excessive smoke and it was because of this that the corporation had approached the government to set a fixed time limit. Given this context, and given the extent to which smoke continued to be a nuisance, polluting the atmosphere and affecting the health of the residents, the Select Committee should have stated the reasoning behind their decision and whether the practical difficulties were of such a magnitude as to destroy the utility of the amended clause altogether. Clearly Mehta did not think so. If the amended section had been passed, prosecutions undertaken and adjudicated in a magistrate's court, he argued, the utility would have been tested. Once prosecuted, the mill owners too would have soon discovered a way to adjust the working of their chimneys so as to reduce the nuisance of smoke and the Municipal Act would have worked as well as it did in England.[78] But the die had been cast and a Bill for the abatement of nuisances arising from the smoke of furnaces in the Town and Island of Bombay was introduced in the Council in June 1911.[79] The government once again outlined the public purpose while securing the support of the business community:

> That there is an urgent need for legislation dealing with this nuisance is admitted on every side. What is visible to the eye cannot be denied; and one has only to approach Bombay by land or sea on a cold-weather morning to find the whole city being gradually enveloped in a dense pall of smoke issuing from hundreds of factory chimneys. The nuisance has been gradually growing until, with the increase in the number of factories and the general substitution of Bengal for English coal, it has now by common consent become intolerable. And Government feel that they have the entire support of the community, not excluding factory owners, in the endeavour to secure a purer and healthier atmosphere for the people to breathe.[80]

In the debate that ensued, two important concerns were raised: (i) regarding the role of the municipal health officer in the proposed commission and indeed the future relation between the commission and the municipality regarding their respective jurisdictions in dealing with smoke, and (ii) regarding the mode of inspection, the desirability or otherwise of the inspector entering private premises without notice for surprise inspections.

Speaking in support of the municipal health officer as a necessary member of the proposed commission was Ibrahim Rahimtoola. His reasoning was straightforward—if the law had failed to curb excessive smoke in the past, it was not because of laxity on part of the health officer but because of lacunae in the law itself. Indeed, this is why greater powers for the municipality were sought in the first place. Now if this was to be denied, it would leave the municipality with little or no knowledge on a rather important issue which continued to be a municipal function.[81] Complaints about the nuisance would continue to be brought to the corporation and nothing but friction and unpleasantness would result from an arrangement in which the municipality was completely excluded.[82] W. D. Sheppard, ex-municipal commissioner of Bombay, stood to counter, arguing that the health officer would best serve his purpose by being an outside critic to the commission when it was formed. The original problem, he argued, had been the difficulty of establishing by evidence the nature of the smoke for which a penalty could be inflicted. The matter was a technical one, and even if the health officer were made a member, he would lack the requisite knowledge about furnaces and machinery, without which he would serve no useful purpose. Even in the past, he had been a poor witness, having to be accompanied for hours at a time by an officer of the maintenance branch, a highly paid mechanical engineer. Nothing could therefore be gained by bringing him back again.[83] Fazulbhoy Currimbhoy concurred. The health officer was not an expert, having failed to curb the smoke nuisance at municipal installations under his watch. But what was more, he went on to add, 'the mill owners are not unmindful of the health of their operatives as the prosperity of their industry depends upon it to a large extent, and they are therefore anxious to improve the surroundings of their operatives and the abatement of smoke nuisance has received their constant attention'.[84] It was left to Sir Pherozshah Mehta to reiterate that the interest of industry did not necessarily coincide with public purpose. The commission to be made, he argued, ought to be composed not only of representatives of one branch of knowledge such as engineering but also people who would be able to give practical and useful advice. And so far as the remedies required for abating smoke nuisance from a public health perspective were concerned, there could not be a better person than the health officer of the municipality to serve on the commission.[85] So the question was posed: was the health officer an expert or not, and if he was one, was he the relevant expert? The government, in its response, both recognised the expertise on offer and denied it a place on the commission:

> A great deal has been said about the expert knowledge of the Medical Officer. That he is an expert as regarding health there can be no question, but as regards smoke he probably does not know much ... The point is to have the best men on the commission, and the question is whether the Health Officer should not be responsible for the inspection of smoke ... it is more advantageous that the Health Officer of the municipality should be an outside observer of the operations of the commission rather than being a member of it.[86]

In another intervention, Rahimtoola proposed that the inspector be allowed to enter the premises without notice. Fazulbhoy responded that this would be a considerable annoyance. But it was left to Shepperd once again to outline just how deeply industrial interests were being prioritised in the discourse regarding public good through the abatement of smoke nuisance. The proposed legislation, Shepperd reasoned, was quite a different thing from the legislation that would have been enforced if the work had been done under the amended Municipal Act. For this was

> a means placed in the hands not of the municipality of Bombay, but of the mill-owners of Bombay, to enable them to rectify in their own mills, by their own methods to be explained to them by their own commission, the nuisance due to their smoke, and this can best be done by the regulations of which the mill owners approve.[87]

If black smoke was emanating from a chimney, the inspector was at liberty to take any notes that he liked but he should not be granted the power to enter the premises for inspection unless the owner had agreed that he should do so.[88] The government, once again, chose to be more partial to this view. While it was necessary to grant some powers to enter private premises in the interests of the community, they reasoned, there was little doubt that this was to some extent also 'an evil'. The law did provide for entry without notice in specific cases, typically having to do with manufacture of coke.[89] But in this instance, the inspector could be allowed to enter only after giving a notice in writing.

Both the amendments having been lost, the Act for Bombay was thus framed on similar lines as that for Calcutta, furnaces being similarly defined and exemption provided for those used for domestic purposes, for burning the dead or for raising steam on ocean-going steamers and on railway locomotives. A Smoke Nuisances Commission consisting of officials and non-officials was provided for, as also the appointment of a chief inspector of smoke nuisances and assistant inspectors to work under the commission.[90] Inspectors were to obtain written permission and give notice before entering for inspection, use and testing of appliances, and directing how furnaces may be worked or stoking done for preventing smoke, without interfering with the business carried on than was absolutely necessary. The government was similarly empowered to prohibit within any specified area the erection of brick, tile or lime kilns, and the erection of furnaces to be used for the smelting of ores or minerals and for the manufacture of coke.[91] Rules were framed prescribing the degree of density of smoke that could be emitted from a furnace, the time during which smoke of such density could be emitted and the altitude below which it may not be emitted.[92] Rules similarly prescribed the period during which, for the purpose of raising steam prior to leaving the port, furnaces of ocean-going steamers would not be considered furnaces within the meaning of the Act, as also prescribe a procedure for giving warnings to offenders before instituting a prosecution. Breach of rules—either by way of smoke in greater density, or at a lower altitude or for a longer time—was to be met with fines of Rs 50 on first conviction, Rs 100 on second conviction and Rs 200 on any subsequent conviction.[93]

The Act for the Abatement of Smoke Nuisances in the Town and Island of Bombay finally came into force on 13 June 1912, jettisoning the proposal for amending the Bombay Municipal Act of 1888 in favour of a law and commission similar to the one in Bengal.

Regulating Smoke: The First Decade

The laws having been framed, it was time to carve out its practical dimensions. William Nicholson, smoke inspector of Sheffield, was appointed as the smoke inspector in Calcutta for an initial period of six months in 1905, before J. Robson and A. Thomson were appointed on a more permanent basis as smoke inspectors, commencing their duties on 1 April 1906.[94] Interestingly, both Nicholson and Robson highlighted the requisite qualities of inspectors and the purpose of inspection along somewhat similar lines. Inspectors needed to be well qualified, even more than engineers, wrote Nicholson, to judge what density and duration of smoke emitted from furnaces and fireplaces were necessary for their working and what were not. The policy for the present, he further suggested, ought to be 'educational' rather than 'adversarial', 'courtesy and co-operation of the mill owners' being critical to the success of the agenda.[95] Robson was similarly inclined, writing that the inspector should not only be a good engineer but also possess tact and sympathy combined with firmness and integrity, so as to put in place an educational policy and win the confidence of owners and magistrates.[96] Both perhaps mirrored the opinion of Angus Smith, the first chief inspector of the Alkali Inspectorate in England, pithily contrasting two styles of inspection: 'One is by a suspicious opponent, desirous of finding evil, and ready to make the most of it. The other is that of a friendly adviser, who treats those whom he visits as gentlemen desirous of doing right', his own preference being for the latter.[97] This attitude was to become even more prominent in the post-war decades, but for now, it stood in combination with the authority to prosecute in the event of smoke exceeding the limits imposed by the rules in the two cities, irrespective of the efforts made by mill owners and managers. In this, there was to be a clear departure from before and immediate success reported too.

In the very first year of its operation in Calcutta, even before the rules had been framed, the commission reported that close to 4,000 observations had been made, 347 tests of appliances done, 71 special appliances fitted, additional boilers introduced in 16 factories and 21 chimneys built, lengthened or reconstructed for the mitigation of smoke.[98] Equally, where close to 8 tonnes of soot were discharged daily at the beginning of 1906 in the oil mill area, only 2 tonnes were being discharged by the end of the year. In Howrah district, 14 minutes of black smoke per hour was reported in the first quarter of the year and only 4.5 minutes in the last quarter. In sum, while dense black smoke was being emitted on an average for 13 minutes in an hour in April 1906, by March the next year, the average had declined to 3.5 minutes. The next year, even further progress was reported, with mechanical stokers and smoke-preventing appliances being fitted to 91 furnaces, statutory warnings for emission of offensive smoke served in 74 cases, directions to increase the height of chimneys issued in 48 cases and prosecutions instituted in three cases, conviction being secured in each instance.[99] The commission received 15 written complaints of smoke nuisances and 'promptly attended' to these, but what was also beginning to get realised was that some of these complaints pertained to the cupolas of iron foundries that were beyond the jurisdiction of the commission, while others included cases of smoke from domestic fires, which too could not be remedied under the Act.[100] On other occasions, much was made of these public complaints, the commission noting that 50 such complaints had been received in 1910–11, as compared to 35 and 22 in the previous two years, suggesting greater public confidence and their appreciation of the relief against smoke that was being experienced.[101]

For the entire next decade, the nature of the reporting on the success of the commission remained of this kind, especially noting the many instances in which height of the chimneys was raised in order to prevent dense smoke from settling near the ground.[102] The number of observations of smoke from chimneys (ranging between 3,500 and 4,000 in various years), inspection of furnaces (2,000–2,500), increased frequency of advice offered to engineers and firemen alike, and a steady growth in the number of statutory warnings that were served to offenders before sanctioning prosecution were other issues reported upon on a regular basis. In 1909–10, 43 warnings were issued, 41 the next year, rising to 72 in 1911–12, before reaching a peak of 86 in 1915–16. Corresponding figures for prosecution in the same years varied from 47 to 51, and any excessive smoke that remained by the end of the first decade of the commission's functioning was entirely attributed to those 'bad factories' that had refused to change.[103] No limit had yet been fixed on the cumulative maximum smoke of density 6 for the city as a whole against which the improvement could be compared each year. And so the most important numbers on offer concerned the steadily declining minutes of average emissions in an hour, starting with a high of 13 minutes when the commission commenced its work, rapidly declining to 2.5 minutes within two years and further declining to 1.31 minutes by 1916–17. Special mention was made regarding success with Bengali management, having persuaded the oil and flour mills of north Calcutta to appoint an engineer to ensure efficient working of their plants.[104] In 1908, a special enquiry was held with reference to the large amount of smoke seen to come from the Howrah side of the river, especially in the evenings during the cold weather, and it was ascertained that though in the aggregate the smoke was considerable, in no individual case did the smoke exceed that permitted by law. Indeed, so confident were the authorities regarding their success in reducing smoke from industrial sources that within three years of the beginning of the work of the commission, they began to make a case for further reduction in the time limit for which dense black smoke could be emitted.[105] It was also proposed to change the definition of furnace to include all types, barring those used in private dwellings for domestic purposes or those used for religious purposes. Open-flame oil lamps that were responsible for much of the smoke in bazaars were also sought to be prohibited, and additional provisions were proposed for making owners and occupiers jointly responsible for coke-making offences. In July 1912, the area under the jurisdiction of the commission was also increased to include the remainder of the limits of Tollygunge, Garden Reach and South Suburban municipalities in the district of 24-Pergunnahs, while a reduced scale of smoke allowance was finally sanctioned in 1913, completely prohibiting black smoke of density 6, and reducing the time limit for other densities of smoke, all as a measure of the confidence of the commission in achieving greater reductions.[106]

A similar style of reporting was adopted in Bombay too, even though warnings and prosecutions here were far fewer than in Calcutta. Premium was still placed on declaring immediate success, with the very first meeting of the commission in Bombay in March 1913 noting reduction in the volume of black smoke as a result of the instructions imparted by Robson to the engineers and firemen in improved methods of stoking and advice to owners and engineers regarding the means to reduce smoke without loss of boiler power.[107] In February 1913, when the commission started its work, the average emission of smoke from each chimney had been the equivalent of 11.5 minutes per hour of dense black smoke. By the end of April the same year, this was believed to have come down to 4.5 minutes and by the end of November to 1.57 minutes. Even more remarkably, all this was reported to have been accomplished without a single prosecution being instituted and with only seven warnings served during the year ending 31 March 1914. The *Times of India* was equally quick to confer success, writing 'we hope good people in Bombay who have enjoyed the coolness of the last few mornings have not failed also to appreciate the thinness of smoke in the air compared with the heavy pall which we have learnt by sad experience to expect as soon as the cool weather begins'.[108] Indeed, such had been the perceived improvement that the commission decided within six months to recommend more stringent limits, being restrained in the matter only because the Act was of recent vintage and, more importantly, had been cordially accepted by the owners.[109]

In the meanwhile, the commission was keen to establish its vigorous approach to complaints received from members of the public or smoke observed by its inspectors, as in Calcutta. In May 1913, it considered a complaint from the Matunga Residents' Association regarding smoke nuisance from the G. I. P. Workshops and Locomotives and from a pottery furnace, serving a statutory warning to the owner of the pottery to discharge the smoke at an altitude of minimum 50 feet.[110] In June, it considered a complaint arising from the engineering workshop of Messrs Richardson and Cruddas, issuing a statutory warning to the firm and giving it three months in which to abate the smoke and to discharge it at an altitude of 100 feet. In September and November, statutory warnings were served to the owners of several other mills to reduce the emission of smoke within 10 days. Alongside, efforts were also made to establish close cooperation with mill owners, with approximately 6,000–7,500 observations being made in the first few years of the commission's functioning and 2,500–3,500 visits undertaken for inspections and training. Most critically the commission seems to have focused a fair degree of its energy on reports and intimations that allowed mill owners and other employers to take action against their employees, especially stokers.[111] In 1914–15, 446 reports and four intimations were sent, 519 reports and six intimations the next year, the numbers rising further to 882 reports and 20 intimations the year after, and in each year, the commission was happy to report that mill owners cooperated fully![112] Such warnings and actions against firemen had featured in Calcutta too—23 offences were reported to the railway authorities in Calcutta in 1907–08, 48 offenders were reported next year with regard to railways and steam launches, reaching even more staggeringly to over 1,000 in 1915–16 and 1916–17, with the commission observing that action taken by owners of factories, steamers, etc., against those in their employ who were responsible for excessive emission of smoke was a critical factor in abating it.[113] Quite clearly, responsibility for excessive smoke was beginning to be split along class lines, passing from managers to workers, even as the Bombay commission wrote that 'the motto of the Commission up to date may be said to be "All is done by kindness"'.[114]

The training of stokers and obtaining possible efficiencies on that count, on the other hand, took some time coming, which is not surprising given that, in general, labour efficiency had not been a priority in either the jute mills of Calcutta or the cotton mills of Bombay.[115] The issue was first raised in Calcutta in 1910, specifically in the context of the difficulties experienced in abating smoke in the northern part of the city that was largely inhabited by Indians who, it was believed, remained rather ignorant about the management of furnaces.[116] Regulations for examination of stokers were approved in 1912 and the commission given power to deal with firemen by instructing and examining them and granting certificates of competency to those who passed the necessary tests. A similar suggestion was made in Bombay too around the same time, with inspectors expected to assist in training stokers for examination of the candidates for certificates of competency to be issued by the commission. Certificates were to be granted in two grades—the lower to be that of 'ordinary stoker' and the higher that of 'tindal stoker', with minimum age of 18 and 20 stipulated for the two categories, respectively.[117] Oral and practical examination were proposed on different aspects of stoking, including peculiarities of different types of furnaces, flues and chimneys, knowledge of stokers' tools and boiler fittings, effect of opening and closing dampers, familiarity with different kinds of coal, the effect of draught on combustion, and the regulation of air supply to suit the rate of combustion and as regards economical prevention of smoke. A fee of Rs 5 was proposed for the examination for an ordinary stoker's certificate and of Rs 8 for a tindal stoker's certificate, and it was mandated that only those with minimum six months and one year experience could be eligible to sit for the exam for ordinary and tindal stokers, respectively. Finally, applicants were required to produce 'satisfactory testimonials as to sobriety and good conduct' for at least the last 12 months' service preceding the date of application.[118]

The proposal was duly approved in April 1914 and a press note issued urging owners to take advantage of the scheme by inducing their employees to appear for the examination.

Three persons took the exam for ordinary stokers in Calcutta in the first year, the number rising to 29 by 1914–15, before falling to nine (plus one tindal) the next year and two (plus one tindal) the year after.[119] Authorities initially suggested that this may have been on account of the unsettled condition of war but soon realised that the demand for such certification, and its potential use in the job market, was probably overestimated to begin with. Fee was accordingly reduced, hoping to attract more candidates, but even then, only two certificates of competency, one ordinary and one tindal, were granted in 1915–16. Bombay was somewhat quicker to recognise this, as no candidate appeared for the exam in the first year even though the commission had taken the trouble to publish a pamphlet titled 'Hints to Firemen on the Management of Boilers' in Marathi, Gujarati and Urdu, priced at 1 anna per copy.[120] Once again, the high fee was seen to be a disincentive, and it was decided to lower the fee to Rs 2 and Re 1, respectively, at any rate as a temporary measure.[121] Even then, no applications for examination were received the next year, though the commission continued to hope that 'mill and factory owners will induce their firemen to qualify and will eventually only employ firemen so qualified'.[122] Calcutta, on the other hand, reported a huge surge in the numbers taking the examination following the reduction in fees—33 in 1917–18 compared to two the previous year—mostly from concerns controlled by Indian owners.[123] Labour efficiency, they continued to argue, would remain 'one of the most effective means of turning coal into active power with a minimum of wastage and smoke'.[124] But success could not be sustained, it would seem, as fines continued to feature more prominently into the next decades. In Bombay, this was to become particularly marked, given the almost complete absence of warnings and prosecutions of managers and owners. No statutory warnings were thus issued and no prosecutions instituted in 1915–16, one statutory warning but no prosecution the year after, and eight statutory warnings issued and one prosecution instituted in 1917, with the engineer being fined Rs 10, figures that stood in sharp contrast to the number of notices sent regarding disciplinary action against stokers.

Notwithstanding, authorities in Calcutta, and to a lesser extent in Bombay, had much to crow about regarding the success of their initiative. Curzon himself would lead in self-congratulation, recollecting in a speech made to the British parliament that when he first initiated a legislation to tackle smoke nuisance, the local government had been quite averse, but in less than three years after the Act had been legislated, the average emission of dense black smoke from the chimneys at Calcutta had diminished by over 80 per cent. Indeed, so successful had they been that Bombay followed suit, with equal success, such that 'at present not a doubt is cast anywhere in India upon the advisability and practicability of this method of dealing with the subject'.[125] The commission in Calcutta was equally keen to take credit, noting the extent to which its success was serving as an exemplar for proposed regulations for Great Britain and Ireland! And even while noting the limitations of the Act, much would be made of the great strides that had been taken:

> The ruler who does something to purify the air of a great city is a real benefactor of his subjects. The city of Calcutta, the second city of the Empire, is under many obligations to Lord Curzon; but one of the greatest of these obligations lies in the fact that Lord Curzon took real steps to purify our air. It is to him that we owe the establishment of the Bengal Smoke-Nuisances Commission ... That Commission has been working for 10 years and has a record of good work of which any commission might be proud.[126]

Indeed, so high was this sense of achievement that by the end of the First World War, the commission in Calcutta could claim that it was reasonable to expect that 'before long Calcutta and its suburbs will be relieved from the inconvenience' altogether.[127]

As with many such claims, it would have been difficult to please all, and reading the commissions' annual reports also calls for some scepticism. Success in obtaining relief from the nuisance would also depend on how magistrates reasoned. In 1910, for instance, one Rameshwar Malia complained against the smoke and noise emanating from the workshops and chimneys of Messrs Jessop and Company as being injurious to the health and physical comfort of the community.[128] This may well have been so, but the magistrate dropped the charge, reasoning that 'there is no nuisance of which the inhabitants of a manufacturing town can reasonably complain, and what nuisance there is likely to be diminished in the near future', echoing arguments made earlier by the municipal commissioner of Calcutta.[129] Equally, the claims of the commission notwithstanding, not all were persuaded about just how much had been achieved. Addressing the First All-India Sanitary Conference in 1911, K. C. Bose, whom we referred to earlier too, spoke at length about the persistence of the smoke in the city:

> I might mention that the smoke nuisance in Calcutta is daily increasing. Every evening long before the sun sets, you will notice a belt of a dark blue coloured smoke forming the boundary of the town, and as soon as the sun goes down the horizon, the girdle or belt, whatever you may be pleased to call it, begins to expand and converts itself into a regular canopy or *shamiana*, as if intended to guard against the access of pure air. This canopy is made up of smoke, presumably emanating from the chimneys of mills and from furnaces of blacksmiths' shops, etc.[130]

Critical comments also featured in the press, with the *Englishman* commenting that while there was a chief inspector of smoke nuisance in Calcutta, he only occasionally prosecuted, with little deterrent effect, and 'the smoke evil' continued to persist. Along the same lines, the *Pioneer* commented that the cover of dust and smoke that settled on everything in the Calcutta winter was as though 'there was not in existence such a thing as a smoke Nuisance Act'.[131] Others highlighted the smoke in specific neighbourhoods—especially Kalighat and Tollygunge areas—on account of the mills located there:

> The whole atmosphere from the further side of Alipore road to the old Steeplechase Course is absolutely surcharged with thick black smoke, belched from dozens of chimneys,

> sooty and pestilential. Everybody sees it, except the smoke nuisance commission; everybody who goes into the neighbourhood suffers from it; and for this reason I suggest that the appropriate course would be to compel the Smoke Nuisances Commission themselves to endure that from which they would not relieve others.[132]

The commission, naturally, begged to differ. A thorough investigation of the conditions of the area, they reported, disclosed a state of affairs that, far from justifying any complaints, proved that the emission of smoke was much below the permissible limit, the result being confirmed by 'several gentlemen in responsible positions connected with the area who had been induced to interest themselves in the matter'.[133] The problem lay with the excessive number of chimneys—150—the total of whose permissible emissions would amount to much the same as if a smaller number of chimneys were emitting dense smoke continuously. In brief, the commission was happy to respond to grievances, and in its own perception had been doing so diligently, but it would be more fruitful if those aggrieved addressed the commission directly, giving the time and place of offence, rather than writing in the press. Even then, a sense of sub-optimal achievement also remained, especially regarding the extent to which the commission had been able to persuade mill owners about why abating smoke mattered:

> Reduction of the emission of smoke means the saving of fuel ... A column of dense black smoke is ... a wastage of fuel ... due either to carelessness or to ignorance. If this fact were more carefully borne in mind by owners of power plant the efforts of the Commission to minimise the pollution of the atmosphere would receive much more enthusiastic support than is at present the case.[134]

In Bombay too many remained dissatisfied with the commission's style of functioning. The *Bombay Chronicle* wrote that while the commission had been keen to claim quick success, to the ordinary observer, Bombay was still as smoky as ever.[135] The *Times of India*, on the other hand, took exception to the style of functioning of the commission, with the work of the commission remaining shrouded in secrecy and the public kept at a hand's distance. 'If so, this is to be regretted', they wrote, 'for the report just issued by the Air Pollution Advisory Board in Manchester shows how extremely interesting to the public this subject can be made.'[136] This was not to be, and even though the commission in Bombay believed that the public showed their confidence and assisted in their work by drawing attention to cases of smoke nuisances that may not have been observed by the inspectors, the numbers do not quite suggest the same, remaining within single digits in the first few years of its functioning.[137] In Calcutta, the commission frankly accepted that it would be impossible to expect the city to be entirely free of smoke, given the large area and the vast diversity of industrial establishments that had to be monitored. There were about 700 factories, large and small, all using inferior Indian coal as fuel. In addition, there were usually between 50 and 60 ocean-going steamers in the port, about 120 inland steam vessels and steam launches on the river and over 200 locomotives plying daily in and out through the locality, again all using coal.[138] Being kept out of the ambit of law, improvement, if any, it was initially believed, could only be obtained through the 'assistance of the owners', by which was implied penal action against stokers.[139] This was to be true of Bombay too, controlling the smoke emitted from railway locomotives and steamers being perceived to be extremely difficult as these were seldom stationary for long.[140] One option was to await the electrification of railways; the other was to rely on fines on stokers, the case being made that it was 'irresponsible' firing that was the cause of such smoke.[141] There was also the issue of smoke from commercial establishments. As the Act currently stood, there was nothing in it to enable action to be taken in respect of the nuisance caused by the smoke emitted from the chimneys and steam engines used for generating electricity, in clubs, hotels and similar institutions.[142] There were about 8,000 trade furnaces in and around Calcutta, the commission noted, the smoke of which under the present rules could be controlled only in so far as the altitude at which it was emitted was concerned. These were furnaces which produced smoke of colours other than grey or black, and the feasibility of altering the rules so as to better control them had been considered, but no decision arrived at yet.[143] The emission of smoke from blacksmith shops, construction works, bakeries, cooking shops, etc., had also been complained about from time to time, and some improvement was made where it had been possible to enforce the use of gas or some other smokeless fuel. But in most cases, difficulties remained with regard to getting the emission of the smoke taken up to a sufficient height, the nuisance not being so much the density of the smoke as the low altitude at which it escaped.[144]

These apart, two issues in particular saw much debate in 1916, especially in Calcutta, the first concerning the two hours' extension granted to ocean-going vessels and the second with respect to the making of coke. We may recollect that Grover had drawn attention both to ocean-going steamers moored on the Calcutta side that could emit low smoke for as long as 20 minutes prior to departure, and to the steam launches whose chimneys were even lower than those on the ocean-going steamers. Nicholson had also drawn attention to the pollution caused during the two hours of statutory 'close time', noting that steamers did not enjoy such statutory 'close time' at any port in Great Britain. The matter was raised on several subsequent occasions too—in 1910, the commission recommended that the exemption should be done away with, only to adopt a more lenient attitude two years later, recommending that a liberal time allowance be given instead so as to bring the smoke from these steamers fully under the control of the commission but without causing hardship to anyone. Owners, perhaps sensing the commission's concern, had begun to take some precautions too, with the average emission of density-6 smoke being brought down to 0.68 minutes by June 1912. However, their interest in the subject seem to have waned soon after, the average rising to 2.85 minutes per hour by September 1914, and even further to 3.18 minutes per hour by September 1915, a condition that the commission felt would deteriorate further if more stringent measures were not taken.[145] This would not be easy, however. At an informal meeting of parties concerned in November 1915, the dilemma regarding smoke from ships was thus laid out—on the one side were the government and the Smoke Nuisances Commission arguing that pollution on account of shipping was rising and needed to be curbed. On the other were the members of the Bengal Chamber of Commerce who believed that concession was required for safety, as in a 'dangerous' river such as the Hooghly, it was absolutely necessary that vessels navigating it must be able to develop maximum engine power at any time.[146] The port commissioners expressed similar views—safe navigation of shipping should be the prime consideration as it involved the safety of the port itself. With the conditions of tide and wind prevailing in Calcutta, it was essential that the last ounce of power that a vessel was capable of developing should be available, and circumstances may arise where the alternative would lie between issuing black smoke and manoeuvring a vessel in a strong current under power insufficient for her safe navigation. The withdrawal of the time of grace, they apprehended, would lead to officers taking undue risk with their vessels, endangering the safety of the port. The Bengal Chamber of Commerce, as also the Calcutta corporation and the port commissioners, was convinced that excessive smoke was not on account of the carelessness of those in charge of vessels but because of the inferior quality of the coal being used. Under the circumstances, members were asked to consider a proposal to reduce the time to half an hour before moving and for an hour after moving, on the assumption that should an amendment be made, owners and agents would cooperate to ensure that the present excessive emission was reduced to reasonable limits, excessive smoke being not only detrimental to public health but also not creditable from a professional point of view! The said amendment was finally discussed in the Council in March 1916 where, based on consultation with experts, bad coal was ruled out as the cause of excessive smoke. Instead, the blame was yet again laid on bad stoking and owners asked to look into the matter. After this, taking 'every circumstance of the case into consideration' and 'considering the interests of the public', the two-hour exemption clause was dropped and the law amended to 'regulate, with due regard to the safety of shipping, the emission of smoke from the furnaces of vessels'.[147]

The second issue concerned coke. As we saw earlier, there had been little unanimity on how to best deal with this. However, recognising that smoke on this count was particularly bothersome in winters, the commission was invested with additional powers in July 1909 for suppressing the making of coke by open fires in the town and suburbs of Calcutta. Matters were reported to have improved initially, with 26 prosecutions undertaken in the first year, and 23 in the next, fines varying from Re 1 to Rs 10.[148] However, ways of evading the Act were also worked out soon enough, including giving false names and addresses, disowning the coal or by moving it to some other place.[149] The municipality also remained opposed to strict actions, and the commission was constrained to observe that 'reasonable relief' had not been obtained.[150] They would need further powers, they argued, making the owner or occupier of the building or the land on which the offence was committed equally liable as the person who was directly responsible. Additional power was also demanded to enable them to seize, confiscate and dispose of the coke recovered. But the proposition met with some opposition. Babu Charu Chandra Sinha argued that such a provision would cause hardship to those owners who had no powers of entry on the lands occupied by their tenants or who had no authority to control the business the latter might carry on. The Calcutta corporation too held the view that it would be unjust to hold an absentee landlord responsible for the offences of his subjects and that the object would be gained if only the person using the premises for the purpose of making coke or permitting its use for that purpose were made liable. Rai Radha Charan Pal Bahadur pointed out that care must be taken that the privacy of women employed in making coal was not violated and moved an amendment along the lines that prior intimation be provided by inspectors before entering any premise occupied by a woman who customarily did not appear in public. The Act, as amended in 1916, kept some of these concerns in mind while ruling that if any person who made coke in or upon any building or land in contravention of the Act, such a person and the owner (if he knowingly permitted the coke to be made) or the occupier of such a building or land would be jointly and severally liable to fine. Coke so made could be seized by an inspector, pending the order of the magistrate, and the inspector allowed to enter premises without notice, except when the premises were occupied by women who needed privacy, in which case prior intimation had to be given to allow them to retreat. The results though, as we will see in Chapter 5, did not always match the ambition!

Sources of smoke apart, two major administrative changes were also brought about through the amendment of the law in 1916. Inspectors were given powers to enter premises without notice and, in a significant revision of the hesitation that had marked offering of technical advice over the years, it was now decided to follow the American model, giving to the commission the power to control the construction of chimneys and flues in order to prevent the erection of unsafe chimneys.[151] Henceforth no furnace, flue or chimney could be erected or re-erected, altered or added to otherwise than in accordance with plans approved by the commission.[152] With this, the commission signalled a move away from being an advisor to mill managers (and an occasional prosecutor) to also being the chief producer of appropriate design and technology. We enquire into this changed orientation in the next chapter. For the moment though, there was a war to contend with, and abating smoke would be, at least temporarily, a secondary concern:

> The military duties of inspectors, the illness of one assistant inspector, and the necessity of exercising the greatest caution in controlling the emission from smoke from factories and other places of smoke emission the output of which is of such supreme importance now, has obviously prevented the commission from achieving all that might have been expected in normal years ... The attention of merchants, factory owners, agents and ship owners etc. has been so absorbed in dealing with unusual business conditions that it has not been possible, nor has it been thought advisable to try, to force on them questions

> relating to smoke control which, had ordinary conditions prevailed, would have been much more vigorously brought to their notice.[153]

Conclusion

The issue of smoke abatement, we have suggested in this chapter, underwent a radical change in the first two decades of the twentieth century, with English expertise drawn upon to fashion a new Act for the abatement of smoke in Calcutta and an American precedent highlighted in Bombay. Key to this transition was the move away from assessments based on the means that had been adopted to curtail smoke to the fixing of time limits for which smoke of various densities could be emitted, such time and density to be recorded more scientifically than before. Education and learning, and a close cooperation between the commissions and industry remained constant, as did the suspicion of native stokers and their disciplining through fines. What was new, however, more in Calcutta than in Bombay, was that this came together with an equal emphasis on prosecution for having emitted excessive smoke. Success was claimed in this regard in both cities, measured through declining averages of dense smoke, counted in terms of average emissions of density-6 smoke, measured in minutes per hour. However, such claims also need to be read with caution, with continued presence of dark smoke and the lack of adequate engagement with members of the public continuing to be highlighted by the press. In time, the Bengal commission itself took note of several shortcomings, especially with respect to the exception granted to shipping in the original Act and with respect to the regulation of smoke that was consequent upon the manufacture of coke. Equally, it signalled a move towards a new way of doing things in the post-war years, taking direct responsibility for advising on technical matters, about which there had been much vacillation till date.

We look at the commissions in their new avatars in the next chapter.

Notes

1. H. Caldwell Lipsett, *Lord Curzon in India, 1898–1903*, London: R. A. Everett & Co., 1903, pp. 8–9.
2. Lipsett, *Curzon*, pp. 96–97, 114.
3. Robin Moore, 'Imperial India', in Andrew Porter, ed., *The Oxford History of the British Empire, Vol. III: The Nineteenth Century*, Oxford: OUP, 1999, pp. 422–46. National efficiency emerged as a major theme in early twentieth-century Britain and in progressive-era USA, containing several different strands. One was 'technocratic': from this viewpoint, all branches of public life, including the practice of government, needed to be organised on scientific lines under the aegis of the relevant 'experts'. Another strand was managerial. The empire, it was said, should be put on a 'business footing' and administered by 'business methods'. See G. R. Searle, 'The Politics of National Efficiency and of War, 1900–1918', in Chris Wrigley, ed., *A Companion to Early Twentieth Century Britain*, Oxford: Blackwell Publishers, 2003, pp. 56–71.
4. Cited in Syed Sirdar Ali Khan, *Lord Curzon's Administration in India: What He Promised, What He Performed*, Bombay: Times Press, 1905, pp. 105–08. At the time of his departure as viceroy, Curzon is reported to have summed up his work in one phrase: '"Efficiency". That has been our gospel. The keynote of our administration.' See T. Raleigh, *Lord Curzon in India*, London: Macmillan and Company, 1906, p. 564. It was another matter though that his ideal of administrative efficiency was seen by many in India more as the monopolisation of power and the establishment of the superiority of the English over Indians. See G. K. Gokhale, *Presidential Address to the Twenty First Session of the Indian National Congress*, Madras: Natesan and Company, 1905, pp. 8–9.
5. Richard Harris, 'Development and Hybridity Made Concrete in the Colonies', *Environment and Planning A*, 40, 2008, pp. 15–36.
6. The trend towards greater professionalisation in scientific matters, however, cannot all be attributed to Curzon's period of viceroyalty. Prior to the setting up of the Board of Scientific Advice, for instance, the Royal Society had already formed an India Advisory Committee in 1898. See Deepak Kumar, ed., 'Colonial Science: A Look at the Indian Experience', in *Science and Empire: Essays in the Indian Context, 1700–1947*, Delhi: Anamika Prakashan, 1991, pp. 6–12.

7. Raleigh, *Curzon in India*, p. 295. The arrival of the expert, however, elicited mixed reactions in the press. Thus on the one hand, the *Pioneer* congratulated Curzon for having initiated the process of curbing smoke. On the other, it opined that compared to the permanent enshrouding of the great manufacturing cities of England, the smoke of Calcutta was but a 'light fleecy cloud' and that it would be foolish to curb smoke at a high cost which would be necessitated on account of the quality of Indian coal and the poor training of stokers. See *The Pioneer*, 6 Mar. 1903; 8 Jun. 1903.
8. The British Indian Association, among others, was of the view that doing so would also be useful for other industrial centres such as Bombay and Cawnpore.
9. I address domestic smoke in Chapter 5.
10. Frederick Grover, *Report on the Abatement of Smoke Nuisance in Calcutta*. NAI, Home, Public B, No. 95, Nov. 1904.
11. Matters were helped though by the fact that the passage of the steamers beyond the limits of Calcutta happened rather quickly.
12. Meanwhile, the corporation was continuing to permit the port commissioners to set up brickfields in the vicinity of Belvedere and Hastings House. See 'Calcutta out of Season', *The Pioneer*, 6 Jun. 1903.
13. Power should be granted to the local government in the proposed legislation, he suggested, to prohibit such activities within certain limits of Calcutta.
14. Matters were not being helped by the fact that in Grover's estimate, Bengal coal was inferior to English coal in steam-raising power by a margin of up to 12–15 per cent generally and extending up to 30 per cent in the case of good Welsh coal.
15. Grover, *Report on the Abatement of Smoke Nuisance in Calcutta*.
16. This was true globally. Among the reasons why businessmen may have ignored the opportunity to save on their coal bills in the USA and Germany, Uekötter suggests the following: lack of information, relatively small expenditure for coal as compared to other running expenses and disinterest of owners in boiler rooms. See Frank Uekötter, 'Divergent Responses to Identical Problems: Businessmen and the Smoke Nuisance in Germany and the United States, 1880–1917', *Business History Review*, 73, 4, 1999, pp. 641–76.
17. This legislation, as we shall see through this book, would, however, always remain cognisant of the fact that industry must be on the same page as the government, and if this was to be so, there

must be genuine savings on fuel costs so as to be able to persuade industrialists of the value of smoke abatement. Recollecting his initiative a decade later, Curzon would underline this fact yet again: 'The real justification for this legislation in the restriction of smoke is not aesthetic but practical, businesslike, and economic. In other words, we can show to these persons that they will be the gainers in escaping the wastage of fuel in which they now indulge by getting a larger return on their outlay.' *Hansard*, 24 Mar. 1914.

18. But he was willing to concede that a legislation may be called for as the amount of smoke would grow in the future. However, if this were to be the case, mill managers must be given liberal opportunities for removing the nuisance where it existed, education and advice, rather than punishment, being the more important factor. NAI, Home, Public, A, No. 221–22, Jun. 1904.
19. NAI, Home, Public, A, No. 221–22, Jun. 1904.
20. NAI, Home, Public, A, No. 221–22, Jun. 1904. The argument that a nuisance was always context specific echoed similar arguments in England. In *Colls versus Home and Colonial Stores Ltd* (1904), Lord Halsbury famously ruled that 'a dweller in towns cannot expect to have as pure air, as free from smoke, smell and noise as if he lived in the country', before going on to add 'and yet an excess of smoke, smell and noise may give a cause of action'.
21. The issue of harassment by those empowered to inspect factories had long figured in debates on factory legislation in India. Thus, at the time of the passing of the first Factory Act in 1881, Government of India had cautioned local governments regarding their choice of inspectors 'for there is always a risk that, in the absence of suitable precautions, administrative zeal in enforcing a law may overstep the limits and prove injurious to the conduct of commercial enterprise'. See J. C. Kydd, *A History of Factory Legislation in India*, Calcutta: University of Calcutta, 1920, pp. 21–22.
22. There were already sufficient safeguards in the Indian Penal Code and in the current Act on smoke nuisance, they argued, to rule out the need for a new legislation.
23. NAI, Home, Public, A, No. 221–22, Jun. 1904. This, the government said, was a false impression and all such sources would be covered in the Act.
24. NAI, Home, Public, A, No. 221–22, Jun. 1904.
25. In their view, therefore, the manufacture of coke must be prohibited within certain limits. NAI, Home, Public, A, No. 221–22, Jun. 1904.

26. NAI, Home, Public, A, No. 221–22, Jun. 1904.
27. NAI, Home, Public, A, No. 221–22, Jun. 1904.
28. Expectedly though, he believed none of these recommendations to be novel, having been already suggested by the Boiler Commission.
29. NAI, Home, Public, A, No. 221–22, Jun. 1904.
30. Grover, *Report on the Abatement of Smoke Nuisance in Calcutta*.
31. Not too long after Grover made his recommendation, scientists in England were developing techniques to analyse the quantity and composition of soot as a way of overcoming some of the difficulties that inhered in a purely visual study of air. One such experiment was carried out in Leeds in 1907. See Peter Thorsheim, *Inventing Pollution: Coal, Smoke, and Culture in Britain since 1800*, Athens, USA: Ohio University Press, 2006, p. 127.
32. The Ringelmann's smoke chart consisted of six squares, corresponding to scales upon a white background. Scale 1 consisted of a white square that had no black line across it and was all white. Scale 2 consisted of a white square with black lines drawn across the white background, vertically and horizontally 1 mm in width and 9 mm apart, the lines increasing in width with each scale, till the square became entirely black in Scale 6. The squares in this scale, if held at a distance of 30–40 paces from the observer, presented the appearance of blacks and greys of varying density with which the colour of the smoke emitted from a furnace could be compared. This was clearly an advance on the earlier methods of observation that had no specified technology, but its benefits might have been limited as this only measured the darkness of smoke, not necessarily the gases that were the most harmful. See Thorsheim, *Inventing Pollution*, pp. 124–25; David Stradling, *Smokestacks and Progressives: Environmentalists, Engineers and Air Quality in America, 1881–1951*, Baltimore: Johns Hopkins University Press, 1999, pp. 72–73.
33. Mention was also made that the inspector must be given a decent salary so as to avoid being corrupted. NAI, Home, Public, A, Oct. 1905.
34. NAI, Home, Public, A, No. 221–22, Jun. 1904.
35. NAI, Home, Public, A, No. 221–22, Jun. 1904.
36. NAI, Home, Public, A, No. 221–22, Jun. 1904. The exchange of views regarding the composition of the commission also offers a glimpse into the continuing racial dimension of the question. This had to do with the presence or otherwise of the police commissioner, who also doubled as the president of the Boiler Commission, on the

commission being proposed. One proposal was for the amalgamation of the Smoke and Boiler Commissions under the second member of the Board of Revenue. But Risley countered this on the grounds that 'Mr. Gupta [presumably the second member in question] has had no dealing with European commercial people, and if we have to choose between him and Mr. Bingell, there can be no question that the latter (notwithstanding his Police association) will be the far more popular and influential.'

37. NAI, Home, Public, A, No. 221–22, Jun. 1904.
38. NAI, Home, Public, A, No. 221–22, Jun. 1904.
39. NAI, Home, Public, B, No. 95, Nov. 1904.
40. WBSA, Judl Progs, No. 16–43, Nov. 1903. Though on other occasions, as we saw in the previous chapter, the Boiler Commission and individual magistrates had been open to offering technical advice.
41. NAI, Home, Public, A, No. 221–22, Jun. 1904.
42. Governments elsewhere too were reluctant to offer specific technical advice, fearing that this may cause them to face liability issues in the event of the failure of a device to produce the desired results. See Uekötter, 'Divergent Responses to Identical Problems'.
43. NAI, Home, Public, A, No. 221–22, Jun. 1904.
44. On this, he was to prove somewhat prescient if we consider the powers assumed by the commissions post–First World War as regards the design of flues and chimenys in their own office to be compulsorily adopted by mill owners (see Chapter 4).
45. Government of Bengal, *Proceedings of the Council of the Lieut.-Governor of Bengal*, Vol. XXXVII, 1905, pp. 5–10.
46. However, the Act specified that it would not apply to already existing kilns.
47. Members of the commission included the commissioner of the Presidency Division as the president, the sanitary engineer, the inspector of factories and the magistrate of Howrah as ex-officio members, and with two members nominated by the Bengal Chamber of Commerce and one member nominated by the Bengal National Chamber of Commerce. The Boiler Commission, evidently, had failed to persuade the government about the need for their presence on the commission.
48. While introducing the Bill, the government took pains to point out that 'a committee, assisted by well qualified Chief Inspector and composed in part of the representatives of the industries is

very unlikely to take hasty measures or to do anything likely to hamper trade'. The rules thus provided that during working hours an inspector could, after giving reasonable notice in writing, enter and inspect an appliance and suggest changes, without hampering business.

49. The initial limits fixed were:

	Density Scale (minutes/hour)			
	6	5	4	Average
Chimney serving 1–2 furnaces	0	2	3	2.21
Chimney serving 3–4 furnaces	0	3	4	3.13
Chimney serving 5–6 furnaces	0	4	5	4.05
Chimney serving 7–9 furnaces	0	6	7	4.97
Chimney serving more than 10 furnaces	0	6	7	5.89

Smoke of the scales of density numbers 1, 2 and 3 could be emitted from furnaces for any length of time. See IOR, V/27/780/15.

50. Government of Bengal, *Proceedings of the Council of Bengal*, 1905, p. 8.
51. 'Bombay Smoke', *ToI*, 2 Jan. 1908.
52. 'Smoky Bombay: A Cheap Preventive', *ToI*, 25 Feb. 1908.
53. *BG*, 30 Sep. 1904; BMC, *Annual Report*, 1906–07, p. 206.
54. 'The Chimney Smoke Nuisance', *ToI*, 9 Nov. 1907.
55. BMC, *Annual Report*, 1907–08, pp. *xiv*, 227.
56. BMC, *Annual Report*, 1908–09, p. 28.
57. BMC, *Annual Report*, 1906–07, p. 206.
58. 'Alleged Smoke Nuisance', *ToI*, 13 Jun. 1906; 'The Smoke Nuisance: An Unsuccessful Prosecution', *ToI*, 20 Jun. 1906.
59. BMC, *Annual Report*, 1908–09, p. 28.
60. To recollect, the issue of abating smoke and fine for the failure to do so had been mentioned in the Bombay Municipal Act in 1872. This was amended in 1888, but neither in 1872 nor in the amended law was a time limit fixed.
61. As an aside, we may also note the health officer's response to a query on whether the emanations from motor vehicles polluted and poisoned the air. Nuisance from the emissions of smoke, he responded, was a preventable one and prohibited under the Bombay Motor Rules. However, it could not be affirmed whether the smell from motor exhausts was injurious to health or not. Also, given the limited number of motors, any poisonous gases given off by them

were so quickly and freely diluted with pure air as to be rendered harmless. See BMC, *Annual Report*, 1906–07, pp. 207–08. On the other hand, car companies sometimes advertised the absence of smoke from their vehicles as an attractive feature. See advertisement for Napier motor cars, *BG*, 22 Nov. 1911. There were occasional fines too, as, for instance, in 1915 when the collector of Bombay reported a car for causing excessive smoke and the driver was fined Rs 15. 'Prosecution under Smoke Nuisance Act', *The Bombay Chronicle* (hereafter *BC*), 11 Dec. 1915. Limited engagement with automobile pollution in Germany and the USA prior to 1945, Uekötter writes, was on account of the failure to establish a definitive link between car exhaust and ill health, such exhaust being seen more as a nuisance than a health hazard. See Frank Uekötter, 'The Merits of the Precautionary Principle: Controlling Automobile Exhausts in Germany and the United States before 1945', in E. Melanie DuPuis, ed., *Smoke and Mirrors: The Politics and Culture of Air Pollution*, New York: New York University Press, 2004, pp. 119–53.

62. BMC, *Annual Report*, 1907–08, p. 227.
63. 'Smoky Bombay: A Cheap Preventive', *ToI*, 25 Feb. 1908.
64. The absence of smoke in the Petit Mill, they pointed out, had been made possible through the use of Broadbent Smoke Preventer, an appliance that was 'simplicity itself' and whose mechanism, being automatic, was not dependent on the reliability, or otherwise, of unskilled labour who tended to shovel in the entire coal at one go rather than smaller amounts more frequently. On other occasions, they mentioned benefits on account of improvements in internal combustion engine. Adoption of such technologies, they wrote, would be more effective than any legislation in creating smoke-free cities. See 'Smoke-free Cities', *ToI*, 27 Feb. 1905.
65. To the extent that equipment that worked well with one kind of coal did not work as well with another, there was some merit to the plea of there being no available technology that could be universally used.
66. 'The Smoke Nuisance', *ToI*, 6 May 1908. Automatic stokers, though much touted as being superior to hand firing, however, belied their name and required close and skilled attention to assure efficient operations. See Stradling, *Smokestacks and Progressives*, p. 77.
67. 'The Smoke Nuisance', *ToI*, 5 May 1908.
68. 'Smoky Bombay: A Cheap Preventive', *ToI*, 25 Feb. 1908. In Europe, they pointed out, reports of boilers routinely contained a

note on furnace draughts. So should it be in Bombay, which when presented in courts would lead to effective prosecution and save subjective judgements. See 'Bombay Smoke', *ToI*, 2 Jan. 1908.

69. BMOA, *Annual Report*, 1908, p. 15. The use of mechanical stokers at the Love Grove pumping station had been abandoned as these were 'ruinously expensive in coal'. The experiment of using liquid fuel at the Colaba air compressor station had also been abandoned 'on account of the cost being very much in excess of coal'. See 'Bombay Smoke Nuisance: Municipal Efforts', *ToI*, 8 Nov. 1906.
70. 'The Smoke Nuisance', *ToI*, 26 Jun. 1908. This view, however, did not go entirely uncontested, Sir Pherozeshah Mehta suggesting that industrialists in Bombay were being disingenuous in delaying things on account of lack of right advice as this could easily be obtained from their peers in Sheffield, etc.
71. Some among the industrialists were also of the view that as the matter concerned mill owners foremost, they should take the lead in bringing out the right kind of expert from England.
72. BMOA, *Annual Report*, 1910, pp. 13–14.
73. Chicago City Council had passed the first smoke ordinance in the USA in 1881. This was amended in 1906–07 to ensure that only qualified engineers could serve as smoke inspectors. These engineers, in turn, relied on instructions to firemen and engineers rather than engaging in court proceedings. See Stradling, *Smokestack and Progressives,* pp. 102–07. Uekötter, by contrast, suggests that the increasing role of smoke inspectors as educators did not imply the lessening of prosecutions. Instead, in Chicago and other cities, the number of suits being filed saw a dramatic rise. See Uekötter, 'Divergent Responses to Identical Problems'.
74. *Bombay Government Gazette* (hereafter *BGG*), 23 Feb. 1911, Part V.
75. *BG,* 23 Feb. 1911.
76. Government of Bombay, *Proceedings of the Council of the Governor of Bombay*, 1911, Vol. XLIX. Bombay: Government Central Press, 1912, pp. 45–47.
77. *Proceedings of the Council of the Governor of Bombay*, 1911, Vol. XLIX. p. 257. The assistance of representatives of mill owners to the Select Committee was justified on the ground that they were the class most directly affected by this legislation. By contrast, Ibrahim Rahimtoola insisted that 'it is not correct to say that the millowners are alone concerned in the proposed amendment of law ... the

whole of the city is more largely concerned in that question', pp. 47–48. Similarly, Sir Pherozeshah Mehta: 'The abatement of smoke nuisance is not in the interest of the mill-owners, as Mr Fazulbhoy says, but it is in the interest of the city. It is not in the interest of the industrial bodies, but it is for the purpose of seeing that these industrial bodies do not affect the health of the city, as they have been doing for years past, that this bill has been brought forward.' *Proceedings of the Council of the Governor of Bombay*, 1912, Vol. L, p. 168.

78. *Proceedings of the Council of the Governor of Bombay*, 1911, Vol. XLIX, p. 259. Mehta also took the opportunity to make it clear that henceforth the municipal corporation could not be held responsible for smoke nuisance in the city.
79. The Bombay Smoke Nuisances Act, 1912, IOR, L/PJ/6 1092, 1912. However, requests for amendment of the Municipal Act so as to empower the municipal commissioner to deal with smoke nuisance from factories continued to be made well into the 1930s. See 'Deputy Mayor for Bombay?', *BC*, 31 Oct. 1935.
80. *Proceedings of the Council of the Governor of Bombay*, 1911, Vol. XLIX, p. 568.
81. In the meetings of the corporation, it had been underlined that it was the duty of the Health Department 'to observe the effect of any trade practice which may be deleterious to the public health, and an excess of smoke from a factory might clearly be deleterious'. Hence the need for the health officer to be on board. On the other hand, some suggested that if the corporation was to be denied powers in favour of a new commission, why should it bother to lend the services of its staff at all. See 'Bombay Corporation: The Smoke Nuisance', *BG*, 23 Jul. 1912.
82. *Proceedings of the Council of the Governor of Bombay*, 1912, Vol. L, pp. 104–05. Rahimtoola pointed to a harmonious style as a special virtue of Bombay style of doing things, arguing that just because no officer of the municipality was on the commission in Calcutta, this was no reason why it should not be so in Bombay.
83. *Proceedings of the Council of the Governor of Bombay*, 1912, Vol. L, pp. 106–07.
84. The question of workers' health is discussed at greater length in Chapter 5.
85. *Proceedings of the Council of the Governor of Bombay*, 1912, Vol. L, p. 108.

86. *Proceedings of the Council of the Governor of Bombay*, 1912, Vol. L, p. 111.
87. *Proceedings of the Council of the Governor of Bombay*, 1912, Vol. L, p. 117.
88. Rahimtoola naturally contested this proposition. The object of the legislation, in his view, was not that the mill owners should be entrusted with its working but that the government had to take the initiative in securing powers to abate the nuisance.
89. *Proceedings of the Council of the Governor of Bombay*, 1912, Vol. L, pp. 118–19.
90. The members of the first Smoke Nuisances Commission included the collector of Bombay as president, chief inspector of steam boilers (ex-officio), two nominees of government, one of whom was the executive engineer to the Bombay municipality, two nominees of the Millowners' Association and one nominee of the Chamber of Commerce, the last two representing 'bodies or associations whose interests are likely to be affected by the Act'. See MSA, General Department, File No. 931, 1913.
91. Originally, naked flame lamps were also to be included among things that could be regulated within specific areas, as these did not consume their own smoke, but the clause was dropped on the advice of the Legislative Department, which opined that this would be 'very indefinite and dangerously wide in its terms as it would seem to cover the case of the common Indian lamp which has naked flame and does not consume its own smoke'. Further, no such clause figured in the Bengal Act, and so there was no reason to introduce it in Bombay either. See IOR, L/PJ/6 1092, 1912.
92. Once again, the details regarding this were to be the most crucial, and were defined in 1913 as follows:

Density Scale (minutes/hour)				
	6	5	4	Average
Chimney serving 1–2 furnaces	1	2	3	3.21
Chimney serving 3–4 furnaces	2	3	4	5.13
Chimney serving 5–6 furnaces	3	4	6	7.05
Chimney serving 7–9 furnaces	4	5	6	8.97
Chimney serving more than 10 furnaces	5	6	7	10.89

In reckoning the amount of smoke emitted by any chimney, classes 4 and 5 were expressed in terms of 6. For this purpose, one minute of class 4 was reckoned as 0.37 of a minute of class 6, and one minute of class 5 as 0.55 of a minute of class 6, that is, if a chimney emitted one minute of smoke of class 4, one of class 5 and one of class 6 in an hour, the emission in terms of class 6 would be 0.37 + 0.55 + 1 minutes, or 1.92 minutes. See IOR, V/27/780/28 and V/27/780/15. Quite clearly, this was a more generous limit than the one set for Calcutta. The rules were soon amended though, and smoke of density 6 entirely banned in Bombay too. MSA, General Department, File No. 809, 1916.

93. Fine for making coke in ovens or with special appliances was fixed at Rs 250 in the first instance, going up to Rs 500 for subsequent offence, while coke made in contravention of this Act but without the use of an oven or special appliances would attract a fine of Rs 25, extending to Rs 50 on subsequent conviction. See *BGG*, 31 Jan. 1912, Part VII.
94. In Bombay, A. H. Fell, who had served as inspector of boilers since 1906, was appointed the first smoke inspector.
95. William Nicholson, *Report on Smoke Nuisances and their Abatement in Calcutta*, Calcutta: Bengal Secretariat Book Depot, 1906. IOR, V/27/780/14.
96. No particular mention was made of race in considering the necessary qualifications of a smoke inspector, but if one looks at the discussion around the appointment of factory inspectors more generally, it would seem that the question of race would have figured in assessing the desirable qualities of a smoke inspector too. See IOR, Q/2/1/179 and Q/2/1/182.
97. Cited in Michael Moran, *The British Regulatory State: High Modernism and Hyper-innovation*, Oxford: OUP, 2003, p. 44. Examining the nature of industrial regulation in Victorian England, Moran writes that its single most common feature was its cooperative character. Though formally based on statute and independent inspectorate, he writes, law was soon marginalised, sanctions became rare and the practice became more about the necessity of trust between the regulator and the regulated. The reason why this should have been so, he further explains, had to do with power relations between owners and inspectors and the disparity of resources between them.
98. Bengal SNC, *Annual Report*, 1906–07.
99. Bengal SNC, *Annual Report*, 1907–08.

100. We do not get many details about the nature of these complaints, but one of these is reported to be from one Mr H. Stephens and others against smoke caused by a gram crushing mill in Kenderdine's Lane in north Calcutta. WBSA, Mun. Progs, B, No. 92–101, Apr. 1908 and No. 156–61, Jun. 1908.
101. Bengal SNC, *Annual Report*, 1909–10. In the next few years, the number of complaints seems to have gone down slightly, being 39 in 1912–13, 40 in 1913–14 and 28 in 1914–15, before rising again to 42 the next year.
102. In the very first year of its operation, the commission reported 21 instances in which this was done, the number going up to 48 the next year, 60 the year after and so on, with owners being made to realise that a better draught, easier steam and smaller consumption of fuel could be obtained by doing so.
103. Fines that were levied varied widely, from as low as Rs 2 to as high as Rs 100.
104. Bengal SNC, *Annual Report*, 1908–09.
105. Bengal SNC, *Annual Report*, 1908–09. Considerable improvement having been achieved, it was decided not to alter the limits as yet. On the other hand, given the feeling of success as regards curbing factory smoke, the commission decided to devote more time to smoke from blacksmith shops, cooking shops, open coke ovens, etc. See 'Smoke-nuisances in Bengal', *The Pioneer*, 30 Sep. 1910.
106. Bengal SNC, *Annual Report*, 1912–13.
107. MSA, General Department, File No. 931, 1913. See also Bombay SNC, *Annual Report*, 1913–14.
108. 'Bombay Smoke', *ToI*, 22 Oct. 1913.
109. However, it also warned that should any chimney continue to emit excessive smoke so as to spoil the effect of the general improvement, the question could be considered again. Bombay SNC, *Annual Report*, 1915–16.
110. The owner of the pottery furnace pleaded that raising the height of the chimney would damage the contents of the kiln. The commission agreed to consider discharge of smoke at 40 feet, instead of 50, but without conceding that any harm would result to the kiln.
111. Grover's observations on Indian stokers had been rather sharp and racially prejudiced, setting the tone for the Smoke Nuisances Commissions that came in the wake of his suggestions.
112. On fines and withholding of wages as strategies of regulating workers in Bombay mills, see Morris D. Morris, *The Emergence of*

an Indian Labour Force in India: A Study of the Bombay Cotton Mills, Bombay: OUP, 1965, especially Ch. 7.

113. Balachandran, in his study of Indian seamen, mentions that in addition to the whip, masters also had their book of penalties, and the articles of engagement of Indian crews laid down a fairly detailed schedule of fines for various misdemeanours. Though he does not specifically mention fines for bad stoking, the larger culture of fines, in addition to physical punishment, is well established. See Gopalan Balachandran, 'Cultures of Protest in Transnational Contexts: Indian Seamen Abroad, 1886–1945', *Transforming Cultures eJournal*, 3, 2, 2008, pp. 45–75.
114. Bombay SNC, *Annual Report*, 1915–16. Anderson similarly concludes that in effect the responsibility for smoke was nearly shifted from employers to workers. See M. R. Anderson, 'The Conquest of Smoke: Legislation and Pollution in Colonial India', in David Arnold and Ramchandra Guha, eds, *Nature, Culture and Imperialism: Essays on Environmental History of South Asia*, Delhi: OUP, 1995, p. 321.
115. Industrial schools for technical training 'to raise the character and position of the working man' had begun to be set up from around the 1870s. However, given the reliance on cheap labour and the peculiarities of the Indian labour market and its modes of recruitment, labour efficiency was not a significant factor in realising profits, and owners in general were not interested in better training of the labour force.
116. Bengal SNC, *Annual Report*, 1910–11. Popular lectures and literature on smokeless combustion of coal were other moves that were designed to secure the same objective.
117. MSA, File No. 895, Box 45, General Department, 1914.
118. BMOA, *Annual Report*, 1914, Appendix I.
119. Bengal SNC, *Annual Report*, various years.
120. Bombay SNC, *Annual Report*, 1916–17.
121. BMOA, *Annual Report*, 1915, p. 19. Matters were not helped by the fact that such examination was not compulsory under the Act.
122. Bombay SNC, *Annual Report*, 1915–16.
123. Bengal SNC, *Annual Report*, 1917–18. The number rose to 43 in 1919, the highest that was to be achieved. The reluctance of British managers to encourage their stokers to obtain such a certificate, however, perplexed the commission, and it hoped that they would soon realise that this would result in saving of the fuel, 'the

conservation of which is of such vital importance now', and also substantial reduction in the pollution by smoke. This would be a good move at any time, they argued, but it was particularly so as a war measure given the necessity of economy regarding fuel consumption.

124. Bengal SNC, *Annual Report*, 1917–18.
125. Speech in British Parliament, *Hansard*, 24 Mar. 1914.
126. Beatson Bell, while moving the Bill for the amendment of the Smoke Nuisances Act in September 1916. IOR, L/PJ/6 1374, 1916.
127. Bengal SNC, *Annual Report*, 1920.
128. 'Smoke and Noise', *ToI*, 12 Nov. 1910.
129. 'Smoke and Noise', *ToI*, 12 Nov. 1910.
130. K. C. Bose, *Tuberculosis*, Appendix 21, Proceedings of the First All-India Sanitary Conference held at Bombay on 13 and 14 November 1911, Calcutta: Superintendent, Government Printing, 1912.
131. 'Calcutta Smoke Nuisance', as reproduced in 'Indian Notes and News', *ToI*, 13 Apr. 1909; 'Calcutta Out of Season', *The Pioneer*, 20 Apr. 1911. According to the *Englishman*, the relative failure of the commission was because it was regarded 'in the light of a natural enemy by manufacturers'.
132. 'Smoke', *The Statesman*, 16 Mar. 1918.
133. Bengal SNC, *Annual Report*, 1917–18. Among those writing in support of the commission was a rice mill owner who appreciated the efforts being made by the commission despite the difficulties that arose from hiring native firemen. In the long run, he wrote, owners like him were convinced that it paid to reduce the smoke, but in the meantime, he could only hope that the commission would not get 'panicky' and instead continue with its current policy, which was more educative than suppressive. 'Smoke at Tollygunge', *The Statesman*, 24 Mar. 1918.
134. Bengal SNC, *Annual Report*, 1917–18.
135. *BC*, 29 Oct. 1915.
136. 'Smoke Abatement', *ToI*, 4 Jun. 1915. It was because such data was not shared with the public that they did not appreciate the true economic or health costs of air pollution, *ToI* wrote.
137. Four to six complaints were reported across various years.
138. Bengal SNC, *Annual Report*, 1910–11.

139. Bengal SNC, *Annual Report*, 1909–10, 1910–11. As mentioned before, smoke from locomotives was seen as being particularly problematic as it was discharged at very low altitudes.
140. Locomotives frequently emitted smoke of density 5 for long periods. However, as the inspector had to observe smoke of this density for nearly six minutes, they nearly always went out of sight before this period had elapsed, except when shunting or standing still. Reducing the observation time to four minutes, the commission believed, would greatly assist in reporting on offending engine drivers. MSA, General Department, File No. 809, 1916.
141. Bombay SNC, *Annual Report*, 1913–14.
142. IOR, L/PJ/6 1374, 1916.
143. Bengal SNC, *Report*, 1913–14.
144. Bengal SNC, *Annual Report*, 1908–09, 1909–10. The difficulty of controlling smoke produced by *bhunjawaalah*s (street snack sellers) was specially mentioned, as also smoke from fire and brick kilns that had been erected prior to the Act having come into operation.
145. Bengal SNC, *Annual Report*, 1914–15.
146. Bengal SNC, *Annual Report*, 1915–16. The number of ships docking at Calcutta over time had increased from 372 in 1833 to 1,126 in 1863 and rose even further with the opening up of the Suez Canal in 1869. A wet dock was built at Khidirpur in the 1890s, and with the outbreak of the First World War, the docks expanded considerably, comprising 18 general berths and 10 coal berths. See Animesh Ray, 'The Calcutta Port', in Sukanta Chaudhuri, ed., *Calcutta: The Living City*, Vol. II, Delhi: OUP, 1990, pp. 123–27.
147. A similar amendment would be made in Bombay in 1920. See IOR, L/E/7/1350.
148. Bengal SNC, *Annual Report*, 1909–10, 1910–11.
149. The number of prosecutions thus went down to six and eight in 1911–12 and 1912–13, before rising again to 30 in 1914–15.
150. The *Times of India* noted that with this, 'coke makers are to be allowed to blacken the air for many years to come'. See 'Smoke in Calcutta', *ToI*, 16 May 1911.
151. A third issue that was raised had to do with the possibility of having more non-official members on the commission 'to lubricate the wheels of administration with a little of the oil of democracy'. However, this was postponed for the moment. See 'The Lower Provinces', *The Pioneer*, 15 Jan. 1916.

152. IOR, L/PJ/6 1374, 1916. Previously the height of such chimneys was frequently prescribed by a municipal officer before the matter came to the notice of the commission, resulting in dual control, which was found objectionable on several counts, particularly by owners who sometimes had to alter their chimney after erection to satisfy the commission's requirements regarding the altitude at which smoke could be emitted.
153. Bengal SNC, *Annual Report*, 1916–17. Matters were not helped either by the fact that good quality coal was being requisitioned by the government for its own needs, compelling mills both in Bombay and in Calcutta to work with inferior coal.

4 Resource Conservation and Scientific Abatement of Smoke, 1917–45

The Commission notes with satisfaction that its efforts, in many respects, are closely akin to the recent endeavours throughout the world to economise fuel, not entirely for the temporary purposes incidental to the war, but chiefly for subsequent peace requirements... This reduction of fuel is of great national, industrial and communal importance. It prolongs the life of the natural resources which can never be replaced, lessens the demand on public services concerned, and as there is a reduction of gases discharged into the atmosphere, there is a corresponding improvement in public health and less damage to public property.[1]

The Commission is gratified at the progress made under the Smoke Act, which primarily was intended for the protection of public health and property, but which to be successful must not be repressive, must secure the cooperation of the owners, the general installation of better furnaces, and finally the scientific utilisation of fuel, which carries with it the highest form of smoke prevention, public protection and fuel economy. This has been the experience of all well-controlled smoke ordinances and Bengal is no exception.[2]

Resource Conservation and Industrial Futures

In May 1916, Government of India appointed an Industrial Commission to examine and report upon the possibilities of further industrial expansion in India. In doing so, it sought to bring together in a coherent manner a set of ideas that had been developing since the turn of the century regarding state intervention to develop both economic capacities (power, Western technology) and human resources (skilled labour, entrepreneurs willing to take risk) in the colonies.[3] The process was far from smooth, however, plagued as it was by contrary desires of promoting a vibrant Indian economy while ensuring that it did not create competition for metropolitan capitalists, maintaining moderate industrial growth in India yet leaving still a vast market for British manufacturers.[4] Equally, it would be one thing for the British to promote industrialism in India and quite another to ensure that this did not drastically transform the agrarian social order on which imperial rule rested.[5] In the long run, thus, industrialism in colonial India remained rather limited. But even with this limited remit, what the debate on industry served to do was to draw attention to the issue of energy, especially the use of coal and electricity within homes and factories, which, in the context of this book, had major implications for how the smoke question was to be tackled in the post-war period.[6]

The case for electric power as the ultimate solution to smoke nuisance had been made quite early in the century, with members of the Boiler Commission in Calcutta observing in 1902 that the smoke nuisance was unlikely to be entirely eradicated so long as coal was used as the source of power in the mills and factories around the city.[7] And even if the possibility of transmitting power on a large scale from generating stations located in the collieries outside Calcutta was as yet 'only a dream of the future', in its realisation lay 'the certain and only drastic remedy for the existing smoke nuisance'.[8] Much the same point was made by Grover, writing that 'the only possible means of absolutely getting rid of smoke' from the factories of Calcutta was by way of transmission of electricity from a distant generating station, preferably situated at the collieries, to the place of its conversion into power. The use of gaseous fuel was also a possibility. A combination of Mond gas producer, gas engines and electrical distribution thus, he too believed, would ultimately solve the smoke problem.[9] Some sign of this development had become evident by the end of the nineteenth century with the Calcutta Electric Supply Corporation (CESC) being established in 1899 and the first big power station set up at Cossipore in 1912, using local coal to generate electricity.[10] The Calcutta mint applied for electricity in 1904, followed by the Bengal Silk Mills the next year. Indeed, Suvobrata Sarkar argues that by the first decade of the twentieth century, the Bengal government placed electrical supply for industrial purposes ahead of the supply of energy for traction, lighting and fans, providing for bulk supply to serve jute mills, cotton mills, paper mills, hydraulic presses, etc.[11] However, pricing remained an issue, evident in the chief inspector of factories' testimony to the Indian Industrial Commission (1916–17):

> It is, moreover, a notorious fact that, in Calcutta where fuel is cheaper than in almost any other city in the world, the rates charged by the Electric Supply Corporation for small power users are so high that the factories have invariably considered it cheaper to install their own small steam plant.[12]

The CESC expectedly refuted the charge, and later reports also testified to the steady growth of its supply in the city, the demand doubling between 1913 and 1919, mainly owing to the growth of the jute mill load.[13] In time, there was demand from Howrah for which cables were laid in the river. Similarly, the operations of the Improvement Trust brought new demands as did the expansion of electric street lighting. Additional power was sought for the docks too, all this, together leading to the sale of electric energy rising from 30 million units in 1918 to 100 million units by 1925.[14] A new southern station was set up in 1926, followed by a brief period of lull before demand again picked up and the annual sale of energy rose from 170 million units to 230 million units or by 35 per cent between 1930 and 1934, largely as a consequence of reduction in the price of electric power for large industrial consumers.[15]

As the work of the Smoke Nuisances Commission got under way in Calcutta, there were thus regular reports reposing faith in the use of electric, oil and gas motors as a permanent remedy for smoke nuisance. The commission itself was in no position to introduce changes in this regard, but its reports served to encourage their adoption on an ever larger scale year after year. In 1907–08, 127 electric motors, two oil motors and one suction gas motor were reported to have been installed in Calcutta.[16] The next year another 146 electric, one oil and six gas motors, cumulatively representing over 1,200 HP, were fitted, many of them replacing steam engines and boilers.[17] Along with this, the use of gas was also extended, special mention being made of its use in brass melting furnaces in 1919, these being the first of their kind in Calcutta, displacing solid fuel fires, which had been the subject of much complaint on account of smoke.[18] Reduction in gas prices that encouraged small factories and mills to substitute it for steam was also appreciated, the commission pointing out that these smaller establishments were especially wasteful regarding fuel use, producing smoke out of proportion to the power generated. The substitution of fuel, they reiterated, was thus beneficial not only economically but also from the perspective of smoke abatement.[19] Indeed, this continued to be the commission's mantra right through the 1920s and 1930s, as it had been since the initial enactment of the smoke laws. Several tests of industrial furnaces by the commission's staff were done in 1921, hoping to enable the adoption of smokeless appliances, while the Oriental Gas Company opened a testing department to address difficulties and doubts regarding the adoption of such appliances.[20] Substantial progress was also made in the electrification of the jute mills and other large industries and the transfer of the Calcutta Tramway Company's load to CESC.[21] The possibility of low-temperature carbonisation was also considered, of the kind being practised in Germany, Britain and the USA, with a view to reducing gas and electricity prices.[22] Prices for small installations not exceeding 4 kW were reduced beginning March 1930, while reduction of rates for large electric consumers was given effect towards the end of 1932, cumulatively resulting in the use of electrical power for industrial purposes growing to 20,000 HP by 1935 and even further to 35,000 HP by 1940.[23] The marvels of gas and electricity were limitless, the commission reiterated time and again:

> Once electricity is introduced in the factory, mill or home it can be used for light, heat and power in a diversity of ways that no other form of energy can equal; its general use will go far to dispel those clouds of smoke that hang over Calcutta during the cold weather evenings.[24]

And they were not alone either, the promise of electricity for clean business also being remarked upon in the press. The substitution of electric motor for steam power, noted the *Amrita Bazar Patrika*, was the only proper remedy for abating smoke. It might mean extra cost for business and increased taxation for the public, but it was the only guarantee for reducing disease and preventing death.[25] The vice-president of the health association of Ward IV, where many of the mills were situated, adopted much the same tone. Prosecutions and paltry fines, he opined, did not abate the nuisance.[26] Electric power was the only remedy, and this being so, licence should not be granted to any factory that did not utilise electric power.[27] Only occasionally would it be noted that the generation of electricity within the city could itself be productive of smoke nuisance:

> In a complaint regarding the smoke, soot and ashes from the chimneys of the Central Power Station owned by the Calcutta Electric Supply Corporation Ltd., no immediate action could be taken as regards the smoke from the heavily over-loaded boilers which, as the owners state, will be brought within the normal limits when the additional boilers now being installed are brought into commission. But, as regards the discharge of grit and cinders, the matter was taken up with the owners, who are voluntarily fitting grit and cinder catchers … hoping to lead to a reduction in discharge of grit and cinders by over 5 tons.[28]

In Bombay too, much was to be expected from electrification. In addition to power for factories, the *Indian Textile Journal* editorialised in September 1899, electric power would 'light our cities, propel our tramcars, pump our sewage, load and unload our ships, propel hundreds of electric motor cars, pull all our *punkhas* and do a large proportion of domestic labor'.[29] A decade or so later, the *Development Report* for the city in 1914 confidently asserted that 'in view of the adoption of electric power and the mitigation of the smoke nuisance', it was not quite imperative to relocate industries, except the offensive trades that could be moved to the extreme north-east of the island.[30] Coal remained important, both Indian coal supplied from Calcutta and that imported from Wales and South Africa, but unlike in Calcutta, hydroelectric power also became available. A licence for setting up a hydroelectric plant at Lonavala was granted to the Tatas in 1906, the government noting several interrelated advantages including cheaper power and better quality of yarn.[31] It noted too that electric power was 'more cleanly and sanitary' and would so 'abate the smoke nuisance'.[32] Coal consumption by the end of the first decade of the century had reportedly reached around 900–1,000 tonnes daily; the Lonavala plant, it was estimated, would reduce this by 320 tonnes per day, supplying approximately 30,000 HP of the total need of 85,000 HP of power.[33] The combustion of this amount of coal had, the Governor of Bombay noted, the effect of vitiating nearly 2,000 million cubic feet of air every day, going on to add that 'the absence of this stupendous daily pollution of the air we breathe would be good for our constitutions, and must go far to restore the lost beauties of Bombay'.[34] Much the same point was made by Sir Dorabji Tata, assuring mill owners about availability of cheap power, savings in cost of machines and boiler room staff, more efficient regulation of work and increased and better outputs. Equally, the city would benefit by getting rid of smoke, thus assuring sanitation and health together with economic gains.[35] And joining the chorus was the *Times of India*, calling the scheme both patriotic and commercial, designed to benefit not an individual concern but all industries in Bombay. It would also, they wrote, serve to reduce the serious and growing evil of smoke nuisance—'the great sanitary evil, poisoning the air and fouling the city with mephitic filth'.[36] On other occasions, though, the optimism was more tempered, especially when it concerned smoke nuisance:

> A good many people look forward with confidence to the inauguration of the Ghat Hydro-Electrical Scheme. If they are wise they will be modest in their expectations. It will be several years before the city draws power from the Ghats, and the maximum estimated supply there from is forty thousand horse-power. More than double that horse-power is already installed in the City, and if it continues to grow at the present rate, the electrical installation will make very little difference to smoke producing capacity.[37]

And again,

> we have several years to wait for the supply of electricity that is to supersede coal in driving our mills and factories, and as the amount of coal burnt on the island of Bombay continues to increase, it is necessary that more effective measures be taken to check the increasing production of smoke.[38]

Some success was to be had in making this transition over the next decade, with a survey on the use of hydroelectric power in India in the 1920s reporting that the cotton mills of Bombay were being increasingly driven by electric power supplied by the Tata scheme, such power being used for industrial furnaces, workshop heating appliances, metal refineries, textile processes, biscuit manufacturers, etc.[39] Railway companies were equally urged to consider electrification, specially of the local services which were a source of great nuisance to the public, and considerable success was reported.[40] By 1928–29, there were 34 fewer steam trains running in and out of Bombay daily on the Great Indian Peninsula Railway line. The numbers were even fewer for the Bombay, Baroda and Central India (BB&CI) Railway, which was reported to have only 20 steam locomotives in use, the number having been reduced by about 40 during the year.[41] In another couple of years, the BB&CI Railway opened a new central station, with all the steam trains starting from here instead of from Colaba, making a considerable difference to the nuisance of smoke in the Fort area. And by the early 1930s, complete electrification of the local trains on both the railways were reported to have been affected.[42] The municipal corporation also entered into agreements for the supply of hydroelectric power for the Love Grove pumping station and the Colaba air compressor stations, even as steam power remained an important back-up arrangement, the type of coal being used for the latter depending on relative price.[43] In part, these transitions may be explained on account of the scarcity of coal or the use of inferior coal whose low thermal value necessitated a larger consumption during and immediately after the war years. This, the commission observed, caused an increase in the amount of coal burnt per square feet of fire grate area per hour, and in mills and factories where boiler capacity was limited, the owners had to resort to improved methods to meet the situation. The prosperity of the industry in Bombay also tempted owners to make large extensions to their mills and considerable difficulty was experienced in obtaining the requisite power. And under such circumstances, some mills took advantage of the power supplied by the Tata Hydro-electric Company, while others took recourse to greater use of oil.[44] In part, it was also a function of relative cost—at 1920 prices if the Bengal coal cost Rs 5 per tonne in the coalfield area, its cost in Bombay was estimated to be about Rs 16 or 17, depending on whether it was brought by train or ship. Moreover, this price was on an upward trend, making it possible for hydroelectric power to compete with coal in a manner that had not been possible in the past.[45] And so, by the end of the decade, the commission was content to observe that the use of electricity and gas for industrial purposes, having steadily increased, there were now healthier habitations and hygienic surroundings to be found in Bombay.[46]

The Indian Industrial Commission, though not directly addressing the question of smoke, also remained hopeful about this transition, noting that the supply of 40,000 HP of hydroelectric power was helping reduce coal consumption in Bombay by about 1,50,000 tonnes per annum or approximately 600 tonnes a day.[47] Its hopeful stance, however, was differently oriented, being focused on resource conservation, especially the conservation of coal, this being a limited national resource. This too was to have a significant bearing on the post-war thinking of the Smoke Nuisances Commissions though. In its review of the sources of power available for India, the Industrial Commission had noted the rather limited prospects of wind power and imported oil, the undesirability of relying on wood for generating steam and even the rather limited growth of hydroelectric power, barring some centres like Bombay and Mysore.[48] On the other hand, it noted the increasing demand for good quality coal, especially in the context of metallurgical developments, making it imperative to introduce economies in methods of mining and consumption.

Around the same time, Government of India commissioned a report for devising the means for reducing the large-scale waste of coal known to occur at the Raniganj and Jharia coalfields.[49] Several reasons were offered for such waste, with roughly a third of the coal being destroyed or lost on account of bad use of pillars, lack of supervision, absence of expert advice and over-extraction given perverse incentives in the form of commissions to the colliery managers.[50] Many solutions were offered to counter this, including the electrification of coal mines, which the government ruled out for the moment.[51] However, there was another proposal on which the government was prepared to break new ground, viz., as regards the setting up of a controlling authority to supervise the negotiation of the lease between the zamindars and their tenants who undertook the actual mining.[52] It was possible that zamindars might consider this an unjustifiable infringement of the rights of private property, arguing that even if there be waste, the laws of property must hold true. Notwithstanding, the government was prepared to step in to regulate:

> Coal is a national asset on which the manufacturing industries and the commercial expansion of the county depend. A landowner or colliery proprietor is at present in a position to waste this national asset without restriction. By such waste he may obtain immediate financial benefit, but he injures the country, damages his property, and diminishes the estate of his heirs. We hold that the state has the right in the interests of the community, to step in and prevent the dissipation of the country's resources. Indian coal is not inexhaustible and scientific mining methods are needed for its conservation and economic extraction.[53]

While the Mining Act had been concerned with the safety of the miner, the proposed legislation regarding the setting up of a Coal Conservation Department would be concerned with 'the safety of the mineral'.[54]

Away from the coalfields, in the city of Calcutta, the Smoke Nuisances Commission was beginning to articulate a similar concern with resource conservation and efficient use.[55] Smoke was a nuisance, and being so, it was an indication that fuel—'a national resource'—was not being efficiently used.[56] Improvements in this regard promised fuel conservation to the extent of 20–30 per cent, which in turn promised multiple goods—prolonging the life of an irreplaceable natural resource, lessening the demand on public services concerned and leading to a reduction of gases discharged into the atmosphere.[57] The argument was to be repeated on numerous other occasions right through the post-war period, with cheap and smokeless power and heat being declared as matters of international importance. 'It appears', wrote the Bengal commission in 1928, that 'we are on the eve of important changes in the use of fuel, which should have far reaching effects, including the gradual passing away of the wasteful use of raw coal, at least in bulk.'[58] There was a new urgency to the question yet again with the onset of the Second World War, the commission observing that on account of the shortage of boiler coal, efficient use of coal was of even more 'vital national importance in these days'.[59] But there was an interesting difference to be noted too, between fining for excessive smoke and pleading for resource conservation, obliquely referred to in the dissenting note of the mining engineer of the Railway Board in the *Report of Coalfields Committee* mentioned earlier. There was, he believed, little to be anxious about the finitude of Indian coal, there being sufficient amounts to serve another 300 years. More crucially, regarding the powers to be assumed by the proposed Coal Conservation Department, he observed: 'in considering waste of coal the consumer with an inefficient steam plant is not only wasting coal but also transport, but it has not yet been suggested that he should be regarded as a criminal for doing so'.[60] So while smoke was in theory open to prosecution, waste would only be subject to the rigours of efficient use.

It would be for the Smoke Nuisances Commission to integrate these issues of resource conservation and smoke abatement, through a renewed focus on efficiency. The efficiency in question had several dimensions—better observation, more efficient firing and good design of furnaces—all being anchored in the new approach adopted by the commissions in the post-war years, viz., to function as a storehouse of technical knowledge whose expertise would be available for all those who sought it.[61] As before, they continued to maintain that they sought to serve all alike, while ensuring that the relief offered did not subject industry to 'unreasonable demands'.[62] What was new post-war, however, was the added refrain that they were entirely disinclined to prosecute any more, wishing to serve solely as advisors and partners rather than as monitors of errant industries. In operational terms, the change was signalled thus: 'smoke prevention, which is so closely allied to industrial effort, should, in the first instance, have been approached from the furnace end of the installation and not from the chimney top', being more about the efficient stoking of fire rather than what went out as waste.[63] Indeed, their name itself was misleading, they suggested, being suggestive of summary repression.[64] Instead, what was desirable was a perception of their work as contributing to 'smoke regulation', which is how it featured internationally!

The Inspectors

The end of the war was also the beginning of a new round of efforts at achieving administrative efficiency through cost-cutting measures. Among other measures suggested was the amalgamation of the Factory, Boiler and Smoke Nuisances Departments in July 1920.[65] The work of the three departments, it was argued, was similar, being concerned with mechanics and steam—the Boiler Inspection Department inspected the boilers and motive power, the Smoke Nuisances Department inspected the chimneys and was concerned with the fuel used in the boilers, and the Factory Department inspected the fencing and working of the machinery put in motion by the boilers and prime movers. In a mill or factory, all three of these branches or departments were placed under a single mill engineer who, in addition to keeping the mill moving with as little smoke through the chimney as possible, was in charge of the mechanics shop and also responsible for the repairs of machinery. There was no reason therefore why in the government too the three units should not be made interchangeable, especially given that the inspectors of the three departments were not very different from mill engineers in their training and background.[66] What was logical from a financial point of view, however, would not necessarily be the most effective means of inspection in the years ahead. Factory reports in Bombay point instead to the increasing number of establishments that had to be inspected with a fewer number of inspectors, resulting in a situation where it was virtually impossible to pay a surprise visit to any concern. They continued to maintain too that the tasks of inspecting factories and inspecting boilers were highly specialised, thus deserving the sole attention of one chief inspector.[67] The commissioners in Bengal were similarly emphatic that given the public health and industrial importance of the work of the department, it was essential that inspectoral staff should be increased:

> In protecting the public and assisting the owners the policy has been to develop the scientific utilisation of fuel … Efficient smoke reduction, having so many phases and being a complicated engineering problem, requires a sufficient and skilled staff. For over two years, the inspectoral staff has been less than the sanctioned strength, and this shortness is being reflected in the work of the department.[68]

What was more, they suggested, the Smoke Department was a paying investment, with the annual recurring saving to industry effected by the department being over 40 times the establishment cost of the department, not to mention the public health benefits that accrued alongside.[69]

The request therefore was for the modernisation of the system, in the form of a central smoke observatory.[70] The government was persuaded, sanctioning the same in Calcutta in 1926, the observatory being situated on the top of a high building from which an eye could be kept on all the chimneys in the control area. Soon, a telephone was also fitted, making possible rapid communication when an offence was observed.[71] Results were once again declared an immediate success, with the number of offences reported to have doubled and a floating number of 40–50 chimneys detected that were always bad. This was not all though, the commission being of the view that while constantly observing smoke from the observatory was essential, it was the inspection and prevention work at the furnaces that finally mattered. The functioning of the commission was also reorganised, with members required to visit different areas once a fortnight with the chief inspector in order to keep in touch with owners and the public. And since members felt incompetent to deal with the many public health questions that were arising, a proposal was also made to sanction the appointment of the director of public health of Bengal to be a member of the commission.[72] The result—a 27 per cent reduction of the daily black smoke by 1930![73] Most importantly, however, a new role was envisaged for inspectors, though with much less success. It was impossible, the commission observed, for inspectors to simultaneously control the area and maintain the necessary contact with installations to effect efficient preventive work so as to reduce smoke. Observing smoke, in any case, was hardly the job for a qualified engineer and combustion specialist tasked with remedying defective installations and advising on fuel economy. It was a job best left to smoke observers, as in any other modern smoke department in the West.[74] Administrative sanction for the same was soon achieved, but fiscal considerations delayed the appointment for a full decade, when finally in 1938 two smoke observers came on board.[75] However with this, another difficulty began, as smoke inspectors increasingly found it difficult to prove the correctness of their records of smoke observation in court![76] In brief, the complaint for much of the 1920s and 1930s was the constraints imposed by the lack of adequate staff, given financial compulsions. And when the staff did become available towards the end of the 1930s, it was difficult to achieve success in courts.

Much the same difficulties were encountered in Bombay too, perhaps to an even greater extent, with the commission gradually moving from monthly meetings to quarterly ones, with the full committee meeting in Bombay twice a year and sub-committees once each in Ahmedabad and Karachi.[77] In the immediate post-war years, there was only one inspector in Bombay to record instances of smoke nuisance, that too intermittently, as he was also required to inspect boilers.[78] In Karachi, the inspection was done by the inspector of factories, boilers and smoke nuisances stationed there who, given constraints of time, could make smoke observations only for a few days each month and not even daily as in Bombay. And in Ahmedabad, shortness of staff prevented inspection from the beginning till November 1923.[79] The question of a full-time smoke nuisance inspector at Karachi did feature in official deliberations in June 1925, but it was turned down on the grounds that the black smoke emissions issued chiefly from chimneys belonging to the North Western Railway and the Karachi Port Trust, which could be easily observed by the boiler inspector who was required to be stationed in Karachi for at least 10 days a month.[80] As in Bengal, the commission emphasised that the inspector's principal duty was the suppression of nuisances and not the mere collection of statistics for purposes of record and comparison, but with little success.[81] And as the duties of the smoke inspector continued to remain tied to the task of boiler inspection, the number of observations being made also showed a constant decline. On other occasions, more local factors came into play, with the decrease in observations at Karachi in 1927–28 attributed to the fact that the majority of factories were closed, and at Ahmedabad the same year, because only five chimneys were observed at a time instead of the mandated ten.[82] The next year when the committee resolved to go back to observation of ten chimneys at a time, the total number of observations still declined as the inspector remained busy with boiler inspection on account of the large number of boilers installed in new mills! Like in Calcutta, a proposal for erecting an observation tower with telephone connection was also mooted for Ahmedabad, only for it to be negatived and instead a cabin on the Madhav Baug high-level water reservoir was proposed to keep an eye on different industries.[83]

Limited number of inspections apart, the responsibility for reducing smoke nuisance and providing the necessary finances for inspection also remained a matter of dispute between municipal authorities and the local government, especially in the wake of the recommendation of the Retrenchment Committee that this cost be borne by the municipalities and other local bodies concerned, as was the case in England. The reasoning for this was simple—it was the duty of municipalities to provide for the abatement of nuisances generally and smoke nuisance in particular, affecting as it did the health and comfort of persons residing within their respective areas.[84] If that be so, municipal representatives in Bombay counter-argued, so must fines accrue to them. That apart, it had been the decisions of the government that had been responsible for some of the smoke in Bombay, and so they must continue to assume responsibility for its abatement too.[85] In Bengal, the debate was to linger even into the next decade, with the Bengal Retrenchment Committee of 1932 again suggesting that as the work of the Smoke Nuisances Commission was confined to Calcutta and its suburbs, so should its costs be met by municipal funds and a fresh legislation be enacted if required to facilitate this.[86] As we shall see below, the Bengal Smoke Nuisances (Amendment Act) of 1923 and Bombay Smoke Nuisances (Amendment) Act of 1929 had already given to the commissions the power to prescribe a scale of fees for the examination and approval of plans, for inspection and testing of instruments and for the grant of permission for the working of furnaces, flues and chimneys, and generally for the services of inspectors.[87] Contribution of municipal bodies to their financial resources would possibly give them further autonomy from those they had to inspect. But this did not quite fructify, and the government continued to be the main source of financing, albeit on a limited scale. Nor did the staff increase in numbers, either at Bombay or at Calcutta, though in the latter instance, the help of additional smoke observers beginning 1938 helped.[88] And with this limited staff, but with possibly a more modernised system of observation and testing, the commissions set out to achieve the most significant goal that had been enunciated through the amendment of 1916, viz., efficiency and greater economy through scientific fuel use, which would also serve the purpose of abating smoke.

Scientific Research and Design Interventions

If limited staff impeded the possibility of fuller realisation of the commission's ambitions, developments in combustion engineering promised far better results than before. The difference, as explained by the Bengal commission, was as follows: In earlier times, combustion appliances were crude and inefficient. The result was that only half of the heat of the coal and often even less was used for making steam, the balance being wasted in smoke and gases. Under such circumstances, reduction of smoke could only be obtained by the greater use of coal, something that owners found unreasonable and oppressive. However, advances in combustion engineering had made boilers better heat-absorbing machines, with 90 per cent of the heat in the coal being transferred to making steam, thereby reducing heat loss and therefore also the difficulty of complying with the smoke requirements.[89] Once again, results in the positive were declared instantaneously, the engagement of combustion engineers and adoption of better methods by one of the large stations in Calcutta reportedly leading to an annual saving in the fuel bill by about Rs 3 lakh, as also to smokeless chimneys.[90] A number of technical experiments being carried out were reported too—the use of preheated air was reported in 1925; the use of steam jets in the ash pits and specially arranged fire bars, leading to reduction in coal use from 50 to 28 tonnes was reported the next year and mention made of the Lancashire boiler installation at a jute mill designed to burn slack coal which, the commission wrote, 'should rank amongst the most economical of its kind in the world, and also … a high achievement in the scientific handling of India's fuel problem'.[91] Indeed, the perception was that so large had been the savings to industry brought about by the commission that one year of the savings, if invested at 4 per cent, would more than cover the cost of the department forever.[92] And there were more savings yet to come, with the commissions keeping in touch with smoke departments and fuel research institutes in Europe and America regarding trends in the scientific use of fuel and latest international achievements in smoke abatement.[93] Smoke may well have been a 'local problem' in Calcutta and Bombay, with their specific atmospheric conditions, commercial and industrial geography and use of inferior coal. But it could not be solved by a 'local view' alone, requiring instead the adaptation of techniques and technologies being developed elsewhere, as for instance low temperature carbonisation being pioneered in Europe and the USA and new apparatus for better combustion developed by the British Smoke Eliminator Company.[94] Similarly, advances regarding the nature of pollution would be based on studies being conducted globally, as, for instance, regarding solids contained in the smoke, a problem that could well be anticipated in Calcutta too.[95] Solutions were also proposed drawing upon the international experience, an instance being the suggestion regarding the installation of suitable appliances for preventing the annual discharge of 20,000 tonnes of solids into Calcutta's atmosphere.[96] The need for efficient grit catchers was similarly emphasised, with 10 grit catchers for coal burning and 130 for light fuel being deployed in the controlled territory in 1928, the annual grit collected being about 15,000 tonnes.[97]

On the other hand, the commission in Calcutta also took pride in the fact that its own work had been much appreciated in the technical press in Europe, which offered the prospect of a reverse transfer of knowledge—methods for smoke abatement developed in India being used internationally.[98] In particular, mention was made of experiments to enable the use of paddy husk as a fuel, especially given high coal prices in the immediate post-war years.[99] There were close to a 100 rice mills in the area controlled by the commission, and it was believed that using husk instead of coal would present no technical difficulty but for the soot and ash that came out of the furnaces and was much complained about by residents in these localities. On the other hand, if the rice mills could not use paddy husk, it was likely that they would have to shut down, with thousands of operatives losing their jobs. There were other economic advantages to be had too—if all the waste products could be used as fuel, it would mean a saving of 80,000 tonnes of coal or about 12–15 lakh of rupees per annum.[100] Within a year of the question being taken up seriously, the commission reported having helped design separators to prevent the discharge of light grit from the chimneys and improved the furnaces such that combustion could be achieved without soot and ash, with close to 40 mills having already installed it. In another couple of years, there was an even greater pride in reporting that the public had benefitted both by less smoke and cheaper rice, while the commission had received requests for advice and assistance in the use of paddy husk as fuel from different parts of the world.[101] Further experiments were made in the coming years, especially regarding the use of exhaust or waste steam from the engine for steaming the paddy, expected to halve the fuel use and also reduce the emission of black smoke. That exhaust could steam the paddy had been already established experimentally, but it appeared that under ordinary conditions, the suspended oil and water contaminated the rice and slowed down production. To address this, therefore, operations were suspended pending the production of a machine that would purify and dry the steam, with Hagan Corporation's exhaust steam purifier eventually proven to do so, earning the gratitude of the local government by way of the award of the Kaisar-i-Hind Medal for Public Service to Robson on his retirement.[102] The commission, it seems, had earned the right to indulge in some hyperbole, writing that the 'object of its endeavours is to render ever improving service to local humanity, which necessitates a continuous passion for developing better means for controlling and handling affairs', and again that 'smoke abatement and cleaning the air we breathe is the service of humanity, than which no higher exists'.[103]

In this broad context of utilising the benefits of combustion engineering, researching or at least keeping abreast of the international research on scientific utilisation of fuel and carrying out local studies to determine the extent of pollution from different sources, the commission also emphasised the need for suitable furnaces and chimneys as having the potential for reducing fuel use by approximately 30 per cent. Chimneys, they explained, performed two functions—first, they created a draught and so facilitated combustion, and second, they helped discharge the product of the combustion at a height at which it would not cause nuisance. For the strictest economy and highest efficiency in steam boiler plants, it was thus absolutely necessary to design a chimney of sufficient height and area to meet all the overload demands while using cheap and inferior quality of coal, as usually obtained in mills and factories. Both defective as well as excessive draught, it was suggested, were equally ruinous to conditions of good efficiency, carrying away the heat that should have been utilised by the boiler and correspondingly reducing the out-turn of the furnaces and the power of the boiler.[104] Dimensions of flues, furnaces and chimneys for the most efficient consumption of coal were thus investigated at length, with a view to avoiding excessive discharge of gases by ensuring the correct mix of air. If there was an insufficiency of draught, the fuel would not burn completely. If too much, the excess air lowered the temperature of the furnace, resulting in not only smoke but also carrying away the heat released from the fuel up the chimney to waste.[105] 'Thus do we see', the commission observed, 'that coal is wasted by too much air; lives are lost by too little pure air.'[106] Training firemen had been important, but as we noted before, had met with limited success—only 11 candidates obtained certificates in 1918–19, nine candidates in 1920, declining further to one in 1921 and two in 1922, numbers which implied that the commission's efforts at enthusing owners in creating a more professional workforce was not quite yielding results.[107]And even if this was so, possibly the plants were of greater importance, 'because with a well-designed plant a good fireman can attain economy and a minimum of smoke, whereas with an unsuitable plant neither is possible'.[108] What was important was to arrive at efficiency of combustion, to be measured through combustion indicators, the commission noting with satisfaction the use of the same by several mills.[109] Thirty-four installations were examined in the first year, 56 plans in 1919, 41 the next year and 89 the year after.[110] Soon though, even this was deemed insufficient, as over 70 per cent of owners failed to produce the quality of plans that could pass muster with the commission, as mandated by law. The commission therefore decided to take charge itself, preparing the plans in its own office for a fee.[111] Prevention, they wrote, was better than cure. Eighty-three plans of installation were examined following this, of which 78 were made in the office of the commission, with much the same trend continuing in the years ahead.[112] And once again, immediate success was advertised, with the number of minutes of black smoke in an hour declining dramatically from 200 to 40 over the course of one season between 1923 and 1924![113]

And yet, instead of the hoped-for decrease in the number of statutory warnings and prosecutions, there had actually been a slight increase. 'Enthusiastic support' was not yet forthcoming for as things stood, 'to most of those touched by such work it is feared that it appears as more of a nuisance than the nuisance it strives to mitigate'. This was especially the case in older plants, where it was practically impossible to prevent excessive smoke. For owners of such mills, smoke abatement could only be looked upon as 'an unreasonable interference with legitimate industrial endeavour', though the commission remained convinced that the more far-sighted view was to make suitable and properly proportioned plants a basic factor in the general policy of running industrial establishments.[114] The issue took an even more urgent turn during the years of the Second World War, with widespread coal shortage necessitating that the power plants and furnaces work at the highest possible efficiency and economy without undue waste of fuel. In a long note on the subject penned in 1944, the commission divided the question of fuel economy into three heads: (i) efficiency of engine plant, (ii) efficiency of boiler plant and (iii) reduction in the waste of power.[115] The first of these required that pistons and valves of cylinders and glands be well adjusted, while the important requirement for the efficiency of boiler plants was to ensure that cold air was not drawn in and that the proportion and sizes of flues were as laid down in rules framed under the Act. The appropriate size of chimneys, as determined by the area of the grate to which they were connected, was recommended too, with further savings in fuel ensured by picking up unburnt fuel from the ashes and through regular inspection, especially if waste products were being used as boiler fuel.[116]

Bombay was to follow in this direction much later than Calcutta, with the Act being amended only in 1929–30 when it was realised that inspectors of smoke nuisances found themselves handicapped in the matter of offering advice, leaving them with no option but to condemn an offending chimney altogether. So, like in Bengal, it was now made obligatory that owners submit all plans for erection, re-erection or alteration of a furnace, flue or chimney to the commission, so that suitable heights, area of flues, etc., could be considered and an accurate and graphic record of the chimneys in the area maintained.[117] However, the numbers involved remained much less than in Bengal, with four plans examined at Bombay and 16 at Ahmedabad in 1931–32; two plans at Bombay, 15 at Ahmedabad and none at Karachi the next year; and 13 at Bombay, two at Ahmedabad and none at Karachi the year after.[118] In part, this may have had to do with the paucity of staff, but in the main, this was because the Bombay Smoke Nuisances Commission kept its major focus on the small industries. We may recall that in the enquiry conducted in 1885, the sanitary commissioner for Bombay had noted that many of the small factories were situated in the most densely populated parts of the city and were 'doubtless a source of much annoyance if not of danger to persons living in their neighbourhood'.[119] Into the next century, petty industries in the city were also called upon by the municipality to erect chimneys over their furnaces to abate the nuisance of smoke complained by the public. In the wake of this, the commission saw as its task the necessity of talking to the owners of these industries and explaining the provisions of the Act to them before initiating any penal action. Realising too that insistence on the standard height would cause considerable trouble and expense to factory owners, it decided to initially follow a lenient policy and permit such industries to erect chimneys below the standard height allowed by the Act, on the clear understanding that these should not prove a nuisance to adjoining houses or premises.[120] Fewer observations in Bombay, 'only taken when necessary for demonstration of methods of firing', it would seem, was the necessary cost of the extra attention given to small chimneys, such as those of railway and road locomotives, of portable and stationary boilers used in engineering construction works and in the upkeep of roads, and of launches and steam ships in the docks.[121] Many drivers and firemen of small boilers were unaware of the existence of the Act and very few of them knew that the emission of dense black smoke was illegal, observed the commission, making it obligatory on their part to pay greater attention to the same right through the 1920s.[122] They were not alone either, with the municipal corporation also undertaking the task of inspection of small workshops and factories that were within their purview. Several of these were fined on account of the smoke or smell nuisance caused by them, but realising that there remained a paucity of space for such factories, many were permitted to carry on their operations in residential premises.[123] Not surprisingly, right through the decade, complaints regarding smoke and other nuisances arising from small foundries, smith shops, etc., continued to pour in, especially in Wards C and E, necessitating further fines and regular inspections.[124]

Small factories also continued to be regularly established, both in the city and the suburbs, some being brought to the notice of the commission by the Bombay municipality, others found out by the smoke inspector. In either instance, the idea remained to offer advice to owners regarding the flues and chimneys to be built instead of prosecuting them. Fifty-six such new factories were reported in 1928–29, including those for metal casting, soap and dyeing factories, oil mills and bakeries, all of which were instructed by the commission. In one of the aluminum factories induced draught fan with air tubes and chimneys was fitted in the casting metal department of the workshop. In other cases, as many as three or four chimneys were installed in one factory to obviate the nuisance as far as possible.[125] The next year, 44 new factories including dyeing, soap and metal-casting factories, oil mills, sugar refineries, bakeries and biscuit manufactories came up, and were similarly advised.[126] A special case was that of the 10 sugar mills situated on Improvement Trust land that were likely to face eviction, though even regarding these, the commission observed, all had complied with the requirements of the Smoke Act.[127] However, on other occasions, notice was taken of the fact that sugar refineries, soap manufacturing, metal-casting and lead-smelting factories that were situated in the residential and business centre of the city continued to be troublesome. The actual density of the smoke emitted from the chimneys of such factories was scarcely above Scale 4, but the problem was their location and its likely impact on public health of those living in the vicinity. On the other hand, earnings of these factories ruled out the possibility of the commission insisting on a large investment on their part, leaving only the possibility of their relocation outside the congested business area.[128] Location apart, the commission also drew atten-tion to the use of rocking fire bars or other improved types of grates, brick arches and baffle plates and to the need for being vigilant as regards economisers.[129] The desirability of using liquid fuel or coal of good quality for abating the smoke nuisance was impressed upon the factory owners of large plants and the use of coke was advised in the case of small industrial concerns.[130] In Ahmedabad, attention was drawn to the need for more boiler power for underpowered mills in which boilers were worked to their utmost capacity and were consequently prolific producers of smoke.[131] Extension and addition was the order of the day in Ahmedabad, noted the commission in 1927, with new engines, new boiler houses, new weaving mills, spinning or weaving plants, night work, etc., but lacking in enough boiler power.[132] Offending mills were thus frequently visited and the owners personally interviewed and requested to adopt effective measures such as the introduction of rocking fire-bars and steam jets in boiler furnaces, or to increase the number of boilers, which would help to reduce the emissions of smoke from their factories.[133]

The amendment to the Act in 1929 brought the focus on small factories yet again, with several establishments expressing ignorance that they were required to take the commission's approval of the plans of proposed furnaces, flues and chimneys.[134] Several rough sketches for the same, suited for small industrial establishments, were thus examined and requisite advice and instructions offered. New institutional arrangements with the municipal authorities were also arrived at, bringing the work of the two institutions closer than in Bengal. Typically, furnaces had been erected in the city without any reference to the commission or compliance with the rules under the Smoke Act, in consequence of which owners were put to additional expense in carrying out the requisite changes. To obviate this, it was now proposed that the owner of a factory who applied for a licence under the

Municipal Act be asked by the department to also comply with the provisions of the Smoke Act before such licence could be granted and the commission simultaneously informed of the application received.[135] On receipt of such information, the factory was visited and instructions were given regarding the requirements of the Act and the opportunity taken to bring to the notice of the owners the necessity of submitting plans, resulting in their gradual familiarisation with the law. Finally, when the suggestions had been acted upon, a recommendation was made to issue the licence, thus preventing duplication of authority and enabling a factory to operate without delay, but without giving up on its responsibility to comply with the needs of smoke abatement.[136] Public complaints regarding smoke nuisances, however, remained, even if steady or declining in numbers—40 in 1934, 29 in 1935 and 35 in 1937, before increasing again to 40 in 1938. And much of this was reportedly against the smaller factories, being especially persistent as regards the smoke from metal-casting *bhutti*es, leading the commission to suggest an amendment of the licence law so as to refuse the use of such premises where there were dwelling overheads for conducting this trade.[137] And when this was not possible, owners were again advised to erect proper hoods and chimneys.[138]

Scientific studies of the atmosphere were also emphasised, the first of these being a study of the constitution and amount of solid matter in suspension in the air in Ahmedabad, carried out by a professor of chemistry, Gujarat College, the study reporting 0.021 gm of solid matter in 1,00,000 cc of air and 72.8 cc of carbonic acid gas in the same volume of air.[139] In Bombay, a more ambitious study was initiated in 1934–35 to 'ascertain by scientific means the extent of air pollution in the city', with Y. G. Naik of the Royal Institute of Science chosen as the expert to carry out the task.[140] Apparatus somewhat similar to those being used in England was utilised to take samples of air at six different places, both in the mornings and evenings, to calculate the number of particles in 1 cc of air. The initial set of observations was taken over seven days in March 1934, reporting on average 2,500 particles of dust or soot for mornings, increasing to 3,300 in the evenings.

The exercise was repeated during the monsoon months, the average now reported declining to 2,000 particles in the morning and 2,500 in evenings.[141] Based on some other calculations, this implied that 17 per cent of the total particles in mornings and 25 per cent in evenings in the non-monsoon months were on account of dust, the remaining percentage being deposits discharged from chimneys.[142] Spatial differences were observed too, with Malabar Hills and the Fort area found almost entirely free of particles, which seemed mostly confined to the industrial areas.[143] A similar study for Ahmedabad was carried out in 1937, which suggested on average the pollution of the city's air to the extent of 3,000 particles per cc, being much higher in mornings with about 8,440 particles, of which 5,440 were on account of mills alone, and again later in the evening, with 14,400 particles, of which 11,400 particles were contributed by mill smoke.[144] While the dust nuisance had been somewhat eradicated by the asphalting of roads, smoke persisted, and it did so to an extent that the municipality expressed the desire to take over the task from the commission, the latter having failed to create a smoke-free city![145] Notwithstanding, much like the Bengal commission earlier, the Bombay commission too would draw attention to the recognition of its achievements abroad, even if the public in India was perhaps not as convinced: 'The general public seldom realise the great value of the work done in inspecting and controlling smoke nuisance. The subject has recently received attention in the United Kingdom where a Government committee under the Chairmanship of Lord Newton recently investigated the problem.'[146] But the true test of their success would lie in the extent to which smoke was actually abated, either through compliance with the recommendations of the commissions or through prosecutions and fines.

PROSECUTION AND 'SUCCESS'

The commissions may have defined their mandate as being 'more advisory than repressive' in the hope of persuading mill owners to reduce smoke to near minimum, but smoke persisted. The consequence was a peculiar mode of reporting in which success was measured each year in terms of the declining minutes of smoke in the hour, while reports on punishment for offences showed a more zigzag path.[147] From this perspective, 1920 was a remarkable year, being the first time since the Act came into force in Calcutta when not a single factory was prosecuted.[148] This was not to last long though, with eight statutory warnings given and nine prosecutions undertaken the next year and 27 warnings and seven prosecutions the year after.[149] But quite clearly, the number of prosecutions was on the decline as the commission developed the means for strengthening cooperation with owners and further strengthening their control regarding construction and reconstruction of furnaces. By 1922, the commission in Calcutta was more than content with its work, writing that the atmosphere of the city was comparatively cleaner than what was usually associated with an industrial and shipping centre, and this despite the fact that the number of mills had doubled since the time of the Act coming into force.[150] The result was that beginning 1925, there would be several years at a stretch when prosecution was not undertaken, though warnings continued to be offered. 'It appears to be the general experience in dealing with the skilled work of efficient smoke prevention', wrote the commission in 1927, 'that a broad educational and cooperative policy gives better results than prosecuting owners *for failing to do that which they do not know how to attempt*.'[151] Modernisation efforts undertaken at the time were reported to have led to significant increase in observations, and concomitant decline in the amount of smoke, with a 27 per cent reduction reported in 1930 alone, bringing down the average emission of each chimney to 0.18 minutes and reducing the daily total to approximately 2,000 minutes, down from 4,000 minutes the previous year.[152] However, this trajectory of 'only advice, no prosecution' was to reverse yet again beginning 1932, as prosecutions were once again resorted to, with 10 in that year and 11 in the next, of which nine were on account of wrongful construction.[153] Fines ranged from Rs 10 to Rs 150, though the commission still remained convinced that if the public followed the advice given by the technical staff, good results could still be achieved without resorting to prosecution. Statutory warnings also continued to be served, and these increasingly came accompanied with the notice that if remedial action was not taken, prosecution would follow, as indeed it did. Twelve prosecutions were reported for 1934, almost all being first-time offenders, with the exception of the superintendent of the Tallah pumping station who was being prosecuted for the fourth time.[154] Once again, fines ranged from Rs 10 to Rs 125, and the commission was left to rue that meagre fines, often less than the inspection fees that would have been payable, not only made a breach of the Act profitable but also tended to encourage some owners to commit such a breach. Numbers again rose dramatically in 1938, with 34 prosecutions, of which 32 led to convictions, but this came attached with the rider that the inspectors were meeting with difficulty in proving the veracity of smoke observations in the court, necessitating the engagement of court inspectors to conduct contested cases.[155] Numbers increased further to 50 the next year and 41 the year after, with impressive conviction rates/compliance, while acquittals were rare.[156] However, the problem of evidence persisted, with smoke inspectors continuing to be assisted by court inspectors and the courts refusing to accept the official records of smoke observation as evidence.[157] As early as 1934, the Bengal commission had exhorted that owners could do more to avoid smoke, 'which would also enable them to avoid too ready (and often unfounded) criticism of the accuracy of the observations made by the inspecting staff of the smoke from their chimneys'.[158] But after the appointment of smoke observers in place of inspectors, even the courts were beginning to cast doubts. The circle had been completed, it would seem. At the turn of the nineteenth century, it was precisely such problems with establishing evidence that had led to the enactment of the law of 1905. Full four decades later, it looked like matters stood not very differently from where they had started, but with one critical difference—the amount of industrial smoke had been steadily reduced in the city, such that the commission could claim major success in regard to the abatement of industrial smoke.[159]

Not all were persuaded though, the *Amrita Bazar Patrika* writing in 1932 that to the people of Calcutta innumerable chimneys continued to 'speak through the dense black clouds of smoke covering their houses, clothes and nostrils', while the *Swasthya Samachar* complained of smoke nuisance in the Maniktola area.[160] Indeed, at stake was the very measure of success, the *Patrika* writing that in the absence of any reliable means of measuring the claims of the commission, the public would continue to rely on ocular evidence rather than on laboratory tests.[161] Incidence of disease and death in mill areas was reportedly higher than in non-mill areas—from March 1931 to April 1932, 79 deaths from tuberculosis were reported in Ward IV, consisting of 10 blocks, half of which were in the mill area that comprised only three of these 10 blocks.[162] And all this because smoke continued to spill from the mills into homes, filling the upper respiratory passages of residents with soot, this being especially aggravated in winters. Indeed, the tenor of the complaint indicated the obverse of what the commission was seeking to narrate—smoke was gradually, but ever so persistently, establishing its stranglehold over Calcutta. And as it did so, it spread terrible diseases of lung and throat, being especially harmful to women.[163] The director of public health wrote of this at length, while drawing attention to its particular geography:

> As the glow in the western sky dies down in the evening, Calcutta becomes covered with a blanket of smoke at night during the winter months and in the spring—particularly in the northern quarters. The southern side of Calcutta is also covered but to a lesser extent. The smoke becomes dense during those evenings and nights when there is no breeze. At times it is so thick that one cannot see anything on the road about a hundred yards away. The thickness of the smoke is so intensive and its effect so impressive that it strikes at once a visitor from outside Calcutta.[164]

The commission, he argued, was doing its bit, but its mandate was limited to thick, black smoke emanated by large factories, leaving unaddressed the smoke emanating from mills situated in dense neighbourhoods, from railways, from motor cars and lorries and, above all, from home fires. They had also, he believed, been possibly too lenient in enforcing the law, being more inclined towards persuasion than punishment. But the commission would stick to its claims, especially regarding the control of the worst kind of smoke, viz., thick, black smoke emanating from the large establishments. This would be so even with the advent of another war. Supplies for war purposes, the commission noted, had led to increased production not only in established industries but also in the setting up of new industries for producing war materials. In the interest of these factories engaged in war production, rules and regulations under the Smoke Nuisances Act could not be universally enforced. But even so, the commission maintained, 'the interests of public health were not ... overlooked'.[165]

Warnings and prosecutions in Bombay seemed to have been an even lower priority, given the focus on observation of smoke, visits to factories and testing of appliances. Things were also more personalised here, the commission instituting the practice of holding conferences with owners whenever offensive smoke was observed rather than issuing written notices. In the case of the railways too, attention to the nuisance caused by locomotives was drawn by personally interviewing the railway authorities.[166] In its very first report, the commission had mentioned the absence of prosecution, the same being the case the year after, with no statutory warnings issued either. Much the same trend continued in the post-war years with notices and intimations being sent, while refraining from prosecution. In 1922–23, 249 notices and four intimations were sent to factories in Bombay, with an additional five notices to locomotives, 14 to steamers and one to steam lorry.[167] The next year, the numbers seem to have doubled with 508 notices, of which 270 were to factories, 48 to locomotives, 13 to steam lorries, 98 to road rollers, cranes, etc., 68 to steam ships, and 11 to tugs and launches. Ten intimations were also sent, all to factories, but no warnings were issued. In Karachi, the same year, four notices were sent, all to factories, one intimation and warning each to a factory and a locomotive, while in Ahmedabad, 110 notices were sent, of which 108 were to factories and one each to locomotive and road roller, as also 37 intimations, all to factories.[168] The year after saw an even further spike in numbers, with over a thousand notices issued in Bombay and 680 in Ahmedabad, but it was only in 1925–26 that we learn of the first prosecutions being instituted against the Gujerath Ginning and Manufacturing Company, the Ahmedabad Ginning and Manufacturing Company and the Raipur Manufacturing Company Ltd in Ahmedabad.[169] The prosecutions were successfully carried through and one more mill—Saraspur Manufacturing Company Ltd—was prosecuted, with each accused being fined Rs 50.[170] Over the next decade and more, successful prosecutions would continue to be rare though. The owners of a smith shop and a coach factory were prosecuted in Bombay in 1930 for failing to erect a chimney within the prescribed time limit.[171] Next year, the owner of a bakery was prosecuted for failing to raise the chimney to the height mandated by law.[172] The managers and engineers of three other mills were similarly fined for excessive smoke in 1934, while the owner of a small dyeing works was charged for the use of furnaces not approved by the commission.[173] Two prosecutions were again instituted the next year, against managers and engineers of the Victoria Mills, Gamdevi, and the Khatau Makanji Spinning and Weaving Mills, Byculla, with the former fined Rs 105 (Rs 15 each for seven consecutive offences), while in the latter instance, the manager and the engineer were fined Rs 5 and Rs 20, respectively.[174] Three prosecutions were also reported for Ahmedabad in 1938–39, to be followed by a major spike the next year when as many as 31 mills were prosecuted, before annual prosecution figures returned to a more normal two to three prosecutions a year.[175] And as for stokers, those appearing for examination continued to figure in negligible numbers, before the regulations relating to the examination of stokers for certificates of competency were eventually cancelled in July 1925.[176] Offending firemen continued to be dealt with 'more effectively' through departmental action, and the commission consistently observed:

> The thanks of the commission are due to the mill and factory owners, heads of engineering departments, public bodies, shipping agencies etc. for their cordial cooperation with the

> Inspectors in securing the abatement of smoke nuisance by dealing departmentally with drivers and firemen found to be careless in causing unnecessary smoke.[177]

Prominent engineers, Indians and Europeans, also joined in the chorus, speaking of 'the waste of coal and the nuisances caused by smoke in our cotton mills and factories not only in Bombay but in the whole of this country by untrained and illiterate Boiler attendants'.[178] But clearly, the combination of departmental actions and technical advice seem to have produced results to the satisfaction of the commission, which noted, for instance, that though there were on an average about 40 ships always in Alexandra, Victoria and Princess Docks, besides tugs, launches, dredgers, hoppers, etc., the special attention paid to them had resulted in smoke being practically reduced to a minimum in the docks.[179]

In sum, the commissions, both in Bombay and in Calcutta, remained committed to their changed orientation, preferring advice on good stoking, careful firing and use of appropriate technology over fines and punishments of managers as a strategy towards smoke abatement. While they seemed to have achieved little success by way of training firemen, they reported with enthusiasm on the technical advances made in cooperation with engineers and owners, declining average emissions of smoke standing testimony to such advances.[180] In Calcutta, smoke that had been observed for 13.1 minutes in an hour in 1906 declined in a secular fashion to 0.08 minutes by 1938, before witnessing a slight increase on account of war conditions to reach 0.12 minutes by 1944.[181] In Bombay, the reported figures were more uneven, declining from 11.6 minutes per hour to 1.144 minutes by 1919, before increasing again to 1.82 minutes by 1922, the reported increase being on account of the use of inferior quality of coal and the massive extension in Bombay mills without accounting for adequate boiler power.[182] From here on, it once again declined to report 0.377 minutes in an hour by 1935, before rising again to 0.645 minutes by 1940, remaining at much the same level till 1949–50.[183] Much the same would be the story in Ahmedabad, with hourly smoke being reduced from 6.56 minutes in March 1924 to

0.99 minutes by 1937, exceeding the 1-minute mark the next year, before declining again, arguably on account of better inspection and because several mills had switched to working with electrical power for their extra load.[184] However, important areas of concern still remained, most importantly as regards domestic smoke, about which the commissions would increasingly observe that it posed an even more serious threat to public health than industrial smoke.

We take up this story in the next chapter.

Conclusion

The abatement of smoke, we note in this chapter, altered in a significant way in the post-war years. Fundamentally, this had to with a recasting of the approach of the commissions, first in Calcutta and then in Bombay, accompanied by amendments in law. When the Smoke Acts had been first enacted, the emphasis had been on advice, and failing this, prosecution, but only if due care had not been taken to limit excessive black smoke. The next step had been to define the time limit, beyond which if smoke continued to be emitted, this would result in fines irrespective of cause, sometimes of managers and engineers, but more often by way of departmental proceedings against stokers.[185] The next phase, which we have described in this chapter, was to move towards a closer involvement with industry by way of offering advice regarding furnace and chimney design, and when these failed to be adopted voluntarily, a new law was brought in that allowed the commissions to make the designs in their own offices. Advice was offered to large and small factories alike, though in Bombay, there seemed to be a far greater investment in the smaller firms that were seen as lacking the requisite technical skills and financial resources to abate smoke. Aiding this new role were developments in combustion engineering that allowed the commissions to pay greater attention to utilisation of power rather than keep their eyes on the waste that was being emitted. The modernisation of smoke observation, with the appointment of smoke observers, contributed too, leaving the inspector, who was a qualified engineer, to attend more to the technical aspects of the issue. The commissions, especially in Calcutta, also kept abreast of major scientific researches in Europe and the USA, reworking these for Indian conditions. In rare instances, they offered their own success as an example, as, for instance, regarding use of paddy husk as fuel. And while all this was for the commissions to do, advantage could also be taken of the initiatives of others, most notably of electricity suppliers and gas companies whose growth, it was felt, would lead to more smokeless fuel being consumed by industries in the two cities. Resource conservation as a national good and as contributing to the bottomlines of industries was also emphasised, and the commissions remained convinced that the two goals could be well aligned with the task of smoke abatement. By the 1920s, both in Calcutta and in Bombay, success was declared as far as the control of industrial smoke was concerned, but perhaps a bit quickly for after going through a phase of 'not prosecution, only advice', the commissions had to again resort to prosecution, with the added burden that courts were reluctant to accept the observations of the smoke observers as evidence, thus returning matters somewhat to as they had been at the turn of the nineteenth century. Not much success was to be had regarding the training of stokers either. On one dimension though, the commission could claim success with some greater degree of certainty, viz., as regards the declining average emissions of smoke in an hour, as emitted by individual chimneys, though segments of the public were far from persuaded that the pall of smoke had indeed been lifted. They also remained consistent in their assurance that complaints from the public would be attended to promptly, and relief given after enquiry, but only to the extent that industry was not subjected to 'unreasonable demands'.[186]

Notes

1. Bengal SNC, *Annual Report*, 1920.
2. Bengal SNC, *Annual Report*, 1920.
3. Clive Dewey, 'The Government of India's "New Industrial Policy", 1900–1925: Formation and Failure', in K. N. Chaudhuri and

Clive J. Dewey, eds, *Economy and Society: Essays in Indian Economic and Social History*, Delhi: OUP, 1979, pp. 215–57.

4. Benjamin Zachariah, *Developing India: An Intellectual and Social History*, Delhi: OUP, 2005.
5. Rajnarayan Chandavarkar, *The Origins of Industrial Capitalism in India: Business Strategies and the Working Classes in Bombay, 1900–1940*, Cambridge: Cambridge University Press, 1994, pp. 3–4.
6. Electricity for lighting purposes could only provide limited revenues to the companies promoting it. So its use for industrial purposes and for transport (for example, electric trams) provided the much needed additional clientele that could attract investors. See Tilman Frasch, 'Tracks in the City: Technology, Mobility and Society in Colonial Rangoon and Singapore', *MAS*, 42, 1, 2012, pp. 97–118.
7. WBSA, Judl Progs, No. 72–76, May 1902.
8. The anticipated displacement of smoke by electricity had also featured in a major way in England from the last decade of the nineteenth century. See Graeme Gooday, *Domesticating Electricity: Technology, Uncertainty and Gender, 1880–1914*, London: Pickering & Chatto, 2008; Peter Thorsheim, *Inventing Pollution: Coal, Smoke, and Culture in Britain since 1800*, Athens, USA: Ohio University Press, 2006, p. 134.
9. Frederick Grover, *Report on the Abatement of Smoke Nuisance in Calcutta*. NAI, Home, Public B, No. 95, Nov. 1904. In this context, it is important to bear in mind that while gas could be clean at the point of consumption, its production was an extremely dirty process, adversely affecting workers and the neighbourhood where production took place. In addition to air, gas works were also highly polluting of water. Use of gas and coke did not eliminate pollution; it simply redistributed it socially and spatially. See Thorsheim, *Inventing Pollution*, Ch. 9.
10. For details regarding electric supply in Calcutta, see Suvobrata Sarkar, 'The Electrification of Colonial Calcutta: Role of the Innovators, Bureaucrats and Foreign Business Organization, 1880–1940', *Studies in History*, 34, 1, 2017, pp. 48–76 and 'Domesticating Electric Power: Growth of Industry, Utilities and Research in Colonial India', *IESHR*, 52, 3, 2015, pp. 357–89.
11. Given the rapid growth in demand from industry, Sarkar writes, CESC decided to install an alternating current plant at Ultadanga station for consumers in the northern parts of the city, the

supply starting in 1910. Sarkar, 'The Electrification of Colonial Calcutta'.

12. Cited in Sarkar, 'Domesticating Electric Power', p. 369.
13. Government of Bengal, *Report of the Committee Appointed to Enquire into the Charges for Electrical Energy Levied by the Calcutta Electric Supply Corporation*, Alipore: Bengal Government Press, 1936, pp. 5–14. IOR, V/26/740/2.
14. Government of Bengal, *Report Electrical Energy*, p. 6. Around this time the feasibility of providing hydroelectric power to the Calcutta mills from water in the Simlipal Hill range was also explored. See GoI, *Electrical Undertakings in India* (corrected up to 1912), Calcutta: Superintendent Government Printing, 1913. IOR, V/27/740/4.
15. Bengal SNC, *Annual Report*, 1932, 1934.
16. Bengal SNC, *Annual Report*, 1908–09.
17. The use of oil fuel on ocean-going steamers, however, could also be problematic if requirements of combustion were not complied with. This was reported to result in dense black smoke, which in its worst stage was heavily charged with oily smuts that caused much damage. Bengal SNC, *Annual Report*, 1919–20.
18. Bengal SNC, *Annual Report*, 1919–1920. This was especially noted as the brass-melting furnaces were located in the congested northern parts of the city.
19. Bengal SNC, *Annual Report*, 1921.
20. Bengal SNC, *Annual Report*, 1921. Details of gas appliances installed for the year mention furnaces for metal smelting, Tinman's stoves, gas engines for power purposes, blowpipes, enamelling stoves and Bunsen burner for lab use.
21. A. T. Cooper, 'Recent Electrical Progress in India', *Journal of the Royal Society of Arts*, 7 Jun. 1929, pp. 739–65. In time, the CESC also appointed its own smoke observer and installed an electric system of smoke indicators at its central power station. Bengal SNC, *Annual Report*, 1927.
22. Bengal SNC, *Annual Report*, 1928.
23. Bengal SNC, *Annual Report*, various years. Around the same time, slack coal began to be used instead of steam coal at the Pulta and Tullah pumping stations in a bid to reduce cost. The effect of this on smoke was not commented upon though. See Calcutta Municipal Corporation, *Annual Report*, 1932–33, 1933–34, 1934–35.

24. Bengal SNC, *Annual Report*, 1931.
25. 'Smoke Nuisance', *ABP*, 12 May 1932. As we note later, the occasion for making this observation was the perceived failure of the Smoke Nuisances Commission in lessening smoke in Calcutta.
26. The owner of 29 Bahir Mirzapore Road Oil Mill, for instance, he pointed out, had been lately fined Rs 125, but the nuisance continued.
27. Existing mills should also be compelled to switch to electric power, he added, and pending such shift they should be asked to use soft coke instead of raw coal. 'Smoke Nuisance in Calcutta: Dr. Sundari Mohan Das's Letter', *ABP*, 12 May 1932.
28. Bengal SNC, *Annual Report*, 1922. Writing of London, Thorsheim points out that as electricity consumption grew in the first decades of the twentieth century, so did the quantity of smoke, with power plants using low quality coal and so producing more smoke and grit than any other source. See Thorsheim, *Inventing Pollution*, p. 135.
29. S. M. Rutnagur, ed., *Electricity in India: Being a History of the Tata Hydro-electric Project with Notes on the Mill Industry in Bombay and the Progress of Electric Drive in Indian Factories*, Bombay: Indian Textiles Journal, 1912, p. 4.
30. *Report of the Bombay Development Committee*, Bombay: Government Central Press, 1914, p. 164.
31. Prior to this, hydroelectric power in India had been generated on a small scale, using natural waterfalls, beginning with the first such plant at Darjeeling in Bengal. The Lonavala plant, by contrast, promised a much larger and constant supply, with power being generated by utilising water from a man-made reservoir.
32. Rutnagur, *Electricity*, p. 11. Much the same point was made by Mr C. H. Armstrong during the course of a public enquiry conducted by the collector regarding the purchase of land for the project in taluka Malwal of Poona district, saying that cheap electric power would be economical for industry as also beneficial from a sanitary point of view, getting rid of smoke nuisance and creating a pure atmosphere. See 'The Tata Scheme: Further Enquiry', *BG*, 7 Nov. 1906.
33. Rutnagur, *Electricity*, pp. 38, 55.
34. Rutnagur, *Electricity*, p. 38.
35. Rutnagur, *Electricity*, p. 30. Within a decade though, critiques of hydroelectric power also began to emerge, noting the economic costs and social impacts of dams, with large tracts of fertile land likely to be submerged and many displaced. Thus when a new plant was

proposed by the Tata company at the confluence of the Mula and Nila rivers in the Mulshi Peta tract of Poona district, there was a long-drawn-out *satyagraha* led by Pandurang Mahadev Bapat between 1920 and 1924, with the overwhelming majority of the protesters being from those likely to lose their land. Rajendra Vora contrasts this with other early anti-dam struggles, such as that against the Grand Canyon dam in the USA, where the campaign was conducted by nature conservationists and not by those directly affected. See Rajendra Vora, *The World's First Anti-dam Movement: The Mulshi Satyagraha 1920–1924*, Delhi: Permanent Black, 2009.

36. *ToI*, cited in Rutnagur, *Electricity*, pp. 46–47. Thus the scheme was referred to by some as the 'true' and 'positive' Swadeshi enterprise as against the 'false' and 'negative' Swadeshi of Bengal, a point that was also made by *Times*, London.
37. 'A Smoke Free Bombay', *ToI*, 22 May 1907.
38. By 1912, there were 30 customers for Tata hydroelectric power, the number rising to 35 by 1916–17, consuming about 45,000 HP. See GoI, *Electrical Undertakings in India* (corrected up to 1916–17), p. 17. By the mid-1920s, hydroelectric power was also being supplied in some mills by the Andhra Valley Power Supply. S. M. Rutnagur, *Bombay Industries: The Cotton Mills*, Bombay: The Indian Textile Journal Limited, 1927.
39. J. W. Meares, *Triennial Report with a Preliminary Forecast of the Water Power Resources of India* (Hydroelectric Survey of India, Vol. III), Calcutta: Government Printing, 1922, pp. 41–42. IOR, V/27/740/11.
40. In cases where a locomotive did not smoke, the soot and grit that were belched out with the exhaust were complained of. See Bombay SNC, *Annual Report*, 1919–20.
41. Bombay SNC, *Annual Report*, 1928–29.
42. Bombay SNC, *Annual Report*, 1930–31, 1932–33. Newspaper reports on the electrification of railways, however, barely mentioned the issue of smoke, focusing instead on other merits of the electrification of railways. See 'Railway Electrification', *BC*, 6 Aug. 1924; 'The New Terminal Station for the City: Governor Opens Bombay Central', *BC*, 19 Dec. 1930. Neither do histories of railways in British India bring up the issue of smoke in any major way. See Ian Kerr, 'Representation and Representations of the Railways of Colonial and Post-colonial South Asia', *MAS*, 37, 2, 2003, pp. 287–326.

43. BMC, *Annual Report*, various years.
44. Bombay SNC, *Annual Report*, 1917–18. Approximately 100 furnaces were reported to have switched to oil. Labour strikes in oil installations in the early 1920s, however, soon forced mills to hurriedly rearrange their furnaces to either partially or wholly burn coal again, and with this, average smoke emission again showed an upward trend. It was apprehended that conversion from coal to oil fuel, especially among smaller factories, might not be sustainable on account of the very low price of coal and the comparatively high price of oil in Bombay. See Bombay SNC, *Annual Report*, 1919–20, 1920–21, 1929–30, 1930–31, 1931–32, 1934–35.
45. The matter of pricing of hydroelectric power was also subject to negotiations between the mills and the power company, especially if the former felt that the cost of hydroelectric power was becoming more than the cost of power supplied by self-contained plants. See BMOA, *Annual Report*, 1922, 1927, 1928.
46. Bombay SNC, *Annual Report*, 1929–30. The number of consumers for electric power in small industries in Bombay was reported to have increased to over 1,200 by 1929. However, things would not change to that degree in other cities. As late as 1939, mill owners in Ahmedabad had failed to utilise electricity in a major way, 'essential for the complete eradication of the nuisance', as they found electricity too costly. 'Smoke Nuisance in the City', *ToI*, 14 Nov. 1939.
47. Indian Industrial Commission, *Inspection Notes 1916–18*, Calcutta, 1918, p. 51. IOR, Q/5/2. When the scheme was fully developed, they expected the supply of power to equal that which would be obtained from the combustion of 4,00,000 tonnes of coal per annum.
48. GoI, *Report of the Indian Industrial Commission 1916–18*, Calcutta: Superintendent Government Printing, 1918, pp. 64–70.
49. GoI, *Report of Coalfields Committee*, Calcutta: Superintendent Government Printing, 1920, p. 1.
50. GoI, *Report of Coalfields.*
51. The prevention of underground waste, they believed, was of far greater practical importance than any saving in fuel consumption. GoI, *Report of Coalfields*, p. 16.
52. In areas of permanent settlement such as Bengal (where the Raniganj and Jharia coalfields were located), the right over minerals was enjoyed by the zamindar. Neither the tenant nor the state had any right over this.

53. GoI, *Report of Coalfields*, p. 5. The need to prevent wasteful use of coal and to conserve it for long-term use figured prominently in England and in the USA too, tied to national business strategies and to imperial concerns. See Stephen Mosley, *The Chimney of the World: A History of Smoke Pollution in Victorian and Edwardian Manchester*, Abindon, Oxon: Routledge, 2008; Ian Tyrell, *Crisis of the Wasteful Nation: Empire and Conservation in Theodore Roosevelt's America*, Chicago: University of Chicago Press, 2015, especially Ch. 5, 'Energy and Empire'.
54. GoI, *Report of Coalfields*, p. 7. Gorman refers to this as 'utilitarian conservation', using resources more efficiently to ensure their longevity. He contrasts this with environmental conservation, which, in his view, includes defining and meeting higher standards of environmental quality. See Hugh Gorman, *Redefining Efficiency: Pollution Concerns, Regulatory Mechanisms and Technological Change in the US Petroleum Industry*, Akron, OH: University of Akron Press, 2001, p. 303.
55. This is where I pick up the story of coal, viz., at the point of consumption. Issues of environmental quality at the mining sites would necessitate another account.
56. Bengal SNC, *Annual Report*, 1918–19.
57. Bengal SNC, *Annual Report*, 1920. Economic anxiety around the exhaustion of coal supplies had figured in England since the 1860s, especially after the publication *The Coal Question* by William Jevons in 1865. Among other things, this thesis pointed out that efficiency gains notwithstanding, the total consumption of coal may yet increase on account of lower prices and new modes of economy, such that efficiency achieved at the level of an individual firm may not translate into systemic gains. On 'Jevon's paradox' and its contemporary implications, see Nuno Luis Madureira, 'The Anxiety of Abundance: William Stanley Jevons and Coal Scarcity in the Nineteenth Century', *Environment and History*, 18, 3, 2012, pp. 395–421; John M. Polimeni, Kozo Mayumi, Mario Giampietro and Blake Alcott, *The Jevons Paradox and the Myth of Resource Efficiency Improvements*, London: Earthscan, 2008.
58. Bengal SNC, *Annual Report*, 1928.
59. Bengal SNC, *Annual Report*, 1944. The issue of fuel economy, in other words, *for purposes of smoke abatement*, moved away from the level of the individual firm to the collective level, though the relationship between the two was not always evident.

60. *Report of Coalfields*, p. 31.
61. Here too the commission may have been mirroring sentiments elsewhere, especially in the USA where Progressive Era anti-smoke movement had gradually toned down its rhetoric. See Robert Dale Gardner, 'From Insurgency to Efficiency: The Smoke Abatement Campaign in Pittsburgh before World War I', *The Western Pennsylvania Historical Magazine*, 61, 3, 1978, pp. 187–202.
62. See, for instance, Bengal SNC, *Annual Report*, 1937, 1941. In her account of expertise among engineers at an oil refinery in Louisiana, Gwen Ottinger makes an interesting distinction between a sense of infallibility as a norm of expertise and another sense of corporate self that projects engineers and scientists as responsible citizens. The radically different national and institutional context notwithstanding, this may be a useful distinction to make in this instance too, with the Smoke Nuisances Commissions rarely presenting themselves as the sole all-knowing experts on the subject but emphasising nevertheless their commitment to public good, serving different interests on the basis of their technical expertise. See Gwen Ottinger, *Refining Expertise: How Responsible Engineers Subvert Environmental Justice Challenges*, New York: NYU Press, 2013.
63. Bengal SNC, *Annual Report*, 1920.
64. Bengal SNC, *Annual Report*, 1918–19.
65. IOR, L/E/7/1106. A proposal for the amalgamation of the Factory and Boiler Departments had been first made by Government of India in 1905 but many provincial governments, Bombay included, had then not thought it fit. However, in 1917, the collector of Bombay changed his view, writing that the smoke nuisance establishment in Bombay should be reduced as the more difficult part of the work had been already done, and if the Act were to be introduced in other cities, a similar path may be adopted. A man of standing and experience would have to be appointed in the first instance to introduce the provisions of the Act. After they have been properly understood, an officer of lesser position and experience could do the work of keeping the emission of smoke within the prescribed limits. If the Smoke Nuisances Department and Boiler and Factory Departments were amalgamated, the government would be able to select any officer they thought fit for the purpose and replace him with another man when the most difficult part of the work had been done. MSA, General Department, File No. 444 of 1920.

66. IOR, L/E/7/1106.
67. Bombay, *Annual Factory Report*, 1920, 1921.
68. Bengal SNC, *Annual Report*, 1921.
69. Lack of staff was also argued to have led to some reversals in the extent of smoke abatement, which having declined steadily up to 1920, went up to 1.6 minutes in 1921 and to 1.68 minutes to an hour the next year. See Bengal SNC, *Annual Report*, 1922, 1925.
70. Bengal SNC, *Annual Report*, 1925.
71. Bengal SNC, *Annual Report*, 1926. Hereafter, observations of factories would be made from the central observatory, while observations of locomotives, steamers, launches and minor industries were made locally. With this, the number of offences detected was reported to have nearly doubled. In Bombay, a sentry box was placed on the Bhandarwada Hill in 1916 to enable the inspectors to take observations of smoke, the position of the hill commanding a comprehensive view of a large number of chimneys. The box also offered shelter to the inspectors from heat and rain, thus enabling them to take the fullest advantage of the position. See MSA, File No. 1056, 1916. In time, more observation points were added. See Bombay SNC, *Annual Report*, 1939–40.
72. We may recollect the extent to which such a proposal had provoked a heated debate about the kind of expertise needed for smoke abatement during the course of the legislative assembly debates in Bombay in 1912.
73. Such declarations of considerable success in post-war years were also common in other parts of the world. Gugliotta mentions that in Pittsburgh, the smoke inspector and the Mellon Institute of Industrial Research claimed to have reduced smoke by 80 per cent between 1914 and 1924. See Angela Gugliotta, 'The "Smoky City" between the Wars', in E. Melanie DuPuis, ed., *Smoke and Mirrors: The Politics and Culture of Air Pollution*, New York: New York University Press, 2004, pp. 100–18.
74. Bengal SNC, *Annual Report*, 1926, 1928.
75. Reports subsequent to this mention that with observers coming on board, the entire area under the control of commission, including the river, was under constant observation in ways that had not been possible before. See Bengal SNC, *Annual Report*, 1940.
76. Bengal SNC, *Annual Report*, 1939.

77. Bombay SNC, *Annual Report*, 1928–29, 1931–32. The provisions of the Smoke Nuisances Act were extended to municipal areas of Karachi in Sind and Ahmedabad in Gujarat from January 1921.
78. Given these conditions, it was decided that all boiler inspectors should have a knowledge of smoke nuisance inspection. Training in this regard would be given to them as and when opportunity arose. Bombay SNC, *Annual Report*, 1922–23.
79. Bombay SNC, *Annual Report*, 1921–22.
80. Bombay SNC, *Annual Report*, 1925–26. The Karachi sub-committee had also proposed that the Act should be amended so as to provide for controlling emission of smoke from oil engines, their reasoning being that the number of oil engines in Karachi, most of them situated amidst living quarters, had increased enormously since 1915. Most of these engines were also observed to be running on crude or refuse oil and, when not properly working, could emit smoke of density 4 and 5, which was even more objectionable to the surrounding neighbourhood than ordinary boilers as the exhaust pipes of these engines were very seldom carried above the surrounding buildings. However, when the issue had been first considered in 1916, it had not merited approval on account of practical difficulties regarding the fixing of a standard, and the Bombay commission did not wish to reopen the matter now either. MSA, File No. 5242-D, 1925.
81. Bombay SNC, *Annual Report*, 1923–24. The inspector took observations of the smoke emitted from the various chimneys indiscriminately, and for the purposes of comparison and record, he also took a series of observations of one hour each of emissions from stationary chimneys from which he complied an average emission for each month and subsequently an average for the year. This average was then taken as representative of the total smoke emissions for the area throughout the year, although the emissions from moveable chimneys did not appear in it. Ordinarily the average emission so obtained was a fairly good index of the amount of smoke in the atmosphere, but 'too great attention should not be paid to sudden fluctuations in the returns', the commission wrote. A decade later, this seems to have changed somewhat with the practice of taking short time observations together with full hourly observations being abandoned. The new system, it was argued, enabled the inspector to devote closer attention to the hourly observations of chimneys

upon which alone were calculated the figures for annual average emission of smoke. Short time observations were also taken but for different purposes and at different times. Bombay SNC, *Annual Report*, 1933–34.

82. Bombay SNC, *Annual Report*, 1927–28.
83. Bombay SNC, *Annual Report*, 1940–41. The proposal had been passed by the Ahmedabad municipality in April 1938. Yet another proposal had been for the appointment of sanitary inspector to observe and report on smoke from mills. See 'Smoke Nuisance in Ahmedabad', *ToI*, 20 Mar. 1938 and 'Preventing Smoke Nuisance', *ToI*, 7 Apr. 1938.
84. *Report of the Bombay Retrenchment Committee*, Poona: Yeravda Prison Press, 1923, pp. 43, 46–47, 94. Savings were estimated at Rs 22,000. The alternative was to recover the cost from mill owners, which is what the commission seems to have favoured.
85. In particular, attention was drawn to the railway workshop between Dadar and Matunga on the land that the corporation had asked for erecting buildings but which the government had instead handed over to the railway, which caused much congestion and smoke.
86. Government of Bengal, *Report of the Bengal Retrenchment Committee, 1932*, Calcutta: Bengal Secretariat Book Depot, 1932, p. 121. However, this did not imply a greater role for the municipalities, as the recommendation was only regarding costs. The government, on its part, accepted the principle that provincial revenues should not be charged with the entire cost of the commission, but it felt that the present time was not opportune to introduce such a legislation. See 'Cost of Bengal Smoke Nuisance Inquiry', *ToI*, 11 May 1934.
87. IOR, L/E/7/1163.
88. In Bombay, by contrast, a smoke observer was added to the staff only as late as February 1947. See Bombay SNC, *Annual Report*, 1946–47.
89. Bengal SNC, *Annual Report*, 1924. Uekötter points out that in theory any improvement in fuel use was bound to lead to better abatement of smoke, but the connection was not made everywhere. In Germany, for instance, unlike the USA, the fight against smoke and the fight for efficient burning remained largely separate issues even after 1914. See Frank Uekötter, *The Age of Smoke: Environmental Policy in Germany and the United States, 1800–1970*, Pittsburgh: University of Pittsburgh Press, 2009, p. 95.

90. Bengal SNC, *Annual Report*, 1926. However, it is difficult to assess to what extent smaller establishments would have benefitted from this transition, given their more limited resources.
91. Bengal SNC, *Annual Report*, 1927.
92. Bengal SNC, *Annual Report*, 1930. The next year the commission also suggested that voluntary contributions must be sought from those who benefitted from the activities of the commission, instead of the entire cost being met by the department.
93. Bengal SNC, *Annual Report*, 1925, 1927. In England, during the First World War, the meteorological office began funding the Committee for the Investigation of Atmospheric Pollution, which measured soot deposits in many of the smokiest cities in Britain. In 1927, this committee was absorbed by the Department of Scientific and Industrial Research and renamed the Atmospheric Pollution Research Committee. See Thorsheim, *Inventing Pollution*, pp. 8, 125–27. In the USA, major smoke-related studies were sponsored by the Mellon Institute, which began support to the University of Pittsburgh's Department of Industrial Research for a comprehensive study of smoke in 1911, issuing nine bulletins in 1913 and 1914. Though the largest such study of the issue yet, scientific research on smoke, however, remained a marginal affair, as elegantly suggested by Harold Platt: 'the battle against urban air pollution established a beachhead of scientific research on a fringe of public health concerns'. See Harold Platt, *Shock Cities: The Environmental Transformation and Reform of Manchester and Chicago*, Chicago: University of Chicago Press, 2005, p. 30. Equally, there was always a tension between ideologies of reform research, industrial research and pure science throughout the period of the Mellon Institute's research on smoke. See Angela Gugliotta, '"Hell with the Lid Taken Off": A Cultural History of Air Pollution—Pittsburgh', Ph.D. diss., Notre Dame University, 2004, pp. 368–84. The results of these studies were also ambiguous, affording evidence both to those who wished to underline success and to those who felt that little had been achieved. See Uekötter, *Age of Smoke*, p. 74; Gugliotta, 'Smoky City', p. 102.
94. Bengal SNC, *Annual Report*, 1929, 1930, 1933. The introduction of air preheaters also drew some appreciation in the press. See 'Purer Air for Calcutta', *The Statesman*, 21 May 1926.
95. Bengal SNC, *Annual Report*, 1927. A small station using 300 tonnes of pulverised coal with 20 per cent of ash daily was estimated to discharge about 17,000 tonnes of grit from the chimney annually.

96. The reason for excessive solids was that the draught was usually increased to lessen smoke and to increase the capacity of the plant, and when this happened, more of the light particles of the fuel were discharged into the atmosphere. Bengal SNC, *Annual Report*, 1925. On this trade-off between increased draught leading to better production and lesser smoke, on the one hand, and the greater escape of light particles into the air, on the other, see 'Purer Air for Calcutta', *The Statesman*, 21 May 1926.
97. Bengal SNC, *Annual Report*, 1928.
98. Bengal SNC, *Annual Report*, 1928.
99. A case of steam paddy husking had first come before the Calcutta High Court in 1918 in which some 10 persons had complained of dust, smoke, noise and smell from the working of one such machine in the Dinajpore municipality, in which the judge had ruled that this constituted a public nuisance under Section 268/290 of the Indian Penal Code. See Calcutta High Court, Bibhuti Bhushan Biswas vs Bhuban Ram on 29 Jul. 1918; (1919) ILR 46 Cal 515. So when the commission decided to take up the issue again, they were well aware that nothing must be done which could result in complaints of public nuisance.
100. Bengal SNC, *Annual Report*, 1921.
101. Bengal SNC, *Annual Report*, 1923.
102. In noting his contribution, the government mentioned that while the daily emission of black smoke in Calcutta had been about 2,00,000 minutes in 1906, this had since reduced to 2,000 minutes by 1930. See Bengal SNC, *Annual Report*, 1929, 1932, 1933.
103. Bengal SNC, *Annual Report*, 1929, 1930.
104. Bengal SNC, *Annual Report*, 1921, 1944.
105. When burning coal, 12–14 per cent of CO_2 was theoretically optimal; if close to 8–9 per cent, then there was a clear loss of efficiency. By the mid-1920s, several of the larger mills had fitted appliances such as thermometers, steam meters and drought gauges for constantly showing how the coal was burned, whether the boilers were operating efficiently and how the steam was being used, and the extent of CO_2. There would no longer be groping in the dark and the rule of thumb would yield to science, the commission observed. See Bengal SNC, *Annual Report*, 1928, 1937.
106. Bengal SNC, *Annual Report*, 1937. During the starting up of the boiler, when the furnaces were relatively cold, some smoke was unavoidable. But once the boilers were running, the correct mix of air

and fuel, it was pointed out, should result in smokeless, and therefore efficient, combustion. 'Smoke Detection', *ToI*, 6 Sep. 1934.

107. Bengal SNC, *Annual Report*, 1918–19, 1920, 1921. In response, the commission considered the preparation of simple directions, in English and in the vernacular, for saving fuel and reducing smoke, with pictures that may 'appeal to the brain through the eye and carry conviction without mental effort', showing, among other things, the losses due to bad firing, which were caused by the gases not being properly mixed and burned, and by the excessive air leaking through thin parts or holes in the fires.
108. Bengal SNC, *Annual Report*, 1920.
109. Bengal SNC, *Annual Report*, 1920.
110. Bengal SNC, *Annual Report*, various years.
111. Bengal SNC, *Annual Report*, 1923. The charging of fees was also as per the recommendations of the Retrenchment Committee.
112. Bengal SNC, *Annual Report*, 1924. Seventy-nine plans were approved in 1925, of which 70 were made in the commission's office, 58 plans were approved in 1926 and 69 in 1927. Plans were also made for mills outside the control area on request. However, the percentage of these that were prepared in the commission's own office seem to fluctuate across the years, with some decline in the 1930s. Of the 52 plans approved in 1931, only 40 were prepared in the commission's office, the corresponding figures for 1934 being 60 and 50, while in 1935, only 63 of the 81 approved plans were made by the commission. By the middle of the 1940s, over a 100 plans were being examined each year. The annual report for 1943–44 reported that 1,951 plans had been approved by the commission since 1916. Bengal SNC, *Annual Report*, various years.
113. Bengal SNC, *Annual Report*, 1924. Improved furnaces and better methods of observing, locating and reporting smoke were offered as reasons for this dramatic improvement.
114. Bengal SNC, *Annual Report*, 1922.
115. Bengal SNC, *Annual Report*, 1944. This was not only a need for individual mill owners but also of vital national importance, the commission reiterated.
116. With bad hand-firing and bad adjustment and arrangements of fire grate in boilers, a large quantity of small pieces of coal and unburnt carbon, the commission observed, was likely to fall down in the ash pit through the fire bars.

117. Bombay SNC, *Annual Report*, 1931–32. The rules reportedly came in for some criticism in the beginning at Bombay, though the commission noted with satisfaction that owners soon came on board, realising the benefits from the point of the efficiency of the boilers and the economic consumption of fuel. However, things were a bit more difficult in Ahmedabad where under-sized chimneys had been first erected in several mills but where trade was increasing to such an extent that extensions had been necessary, and consequently the flue and chimney areas had been found to be too small for the larger plant.
118. The maximum that the Bombay commission seems to have examined in a year seems to be 29 plans for Bombay in 1938–39. In many cases, rough sketches were prepared in the office of the chief inspector and supplied to industries to assist them in the proper building of flues and chimneys. Bombay SNC, *Annual Report*, 1931–32.
119. *Report and Proceedings of the Commission Appointed to Consider the Working of Factories in the Bombay Presidency*, Bombay: Government Central Press, 1885, p. 201. IOR, V/26/670/86.
120. Bombay SNC, *Annual Report*, 1918–19.
121. Bombay SNC, *Annual Report*, 1929–30.
122. This assumed even greater centrality during periods of mill strikes such as that between 15 September and 2 December 1925, when the inspector devoted his time exclusively to the abatement of smoke nuisances. See Bombay SNC, *Annual Report*, 1925–26.
123. BMC, *Annual Report*, 1919–1920. Fifty-five cases relating to smoke and other nuisances were dealt with during 1924–25, 38 cases the next year and 36 cases in 1929–30.
124. BMC, *Annual Report*, 1923–24.
125. Bombay SNC, *Annual Report*, 1928–29.
126. Despite the continued depression in trade and its consequences on industries generally, furnaces continued to be erected all over Bombay in connection with the smaller industries such as metal casting, those dealing with precious metals and manufacture of soap. Bombay SNC, *Annual Report*, 1931–32.
127. Bombay SNC, *Annual Report*, 1929–30.
128. Bombay SNC, *Annual Report*, 1930–31.
129. Bombay SNC, *Annual Report*, 1919–20. In many factories, economiser tubes were allowed to remain in use till they were nearly

choked up with scale, thus rendering them useless. Noting this, the commission advised that the value of an economiser lay in its capacity to efficiently heat the feed water and thus reduce coal consumption, but as soon as scale started forming on the inside of the tubes, its value decreased.

130. Bombay SNC, *Annual Report*, 1934–35. The use of coke as a fuel in smaller establishments had been around for a decade and more, but it was pointed out that in addition to its smokeless quality, use would also depend upon the extent to which it cost or provided greater convenience or, even preferably, both. See 'Low Temperature Coke and Smoke Abatement', *ToI*, 4 Jul. 1924.
131. Bombay SNC, *Annual Report*, 1923–24, 1924–25. Very few mill owners in Ahmedabad were reported to have a reserve of boiler power and most were unable to meet the demand for extra steam, especially in the cold weather, without forcing their boiler plant to work at excessive capacity. Another factor contributing to excessive smoke in Ahmedabad was the high cost of good coal.
132. Bombay SNC, *Annual Report*, 1927.
133. Bombay SNC, *Annual Report*, 1929–30.
134. Bombay Act X of 1920 mandated such compulsory approval of plans for erection or alteration of furnaces, flues and chimneys, but it would be a few years before the commission was confident that smaller owners were beginning to submit copies of plans for proposed furnaces and chimneys. Bombay SNC, *Annual Report*, 1932–33.
135. Licence conditions to ensure smokeless combustion had been first introduced in Prussia in 1853 and became one of the major pillars of the smoke law across Germany by 1900. In the USA, Chicago was the first city to refuse permits for new fuel-burning plants unless approved by the city. See Frank Uekötter, 'Divergent Responses to Identical Problems: Businessmen and the Smoke Nuisance in Germany and the United States, 1880–1917', *Business History Review*, 73, 4, 1999, pp. 641–76; Ian Douglas, *Cities: An Environmental History*, London: I. B. Tauris, 2013, pp. 77, 79.
136. Cooperation of the municipality was also sought in Ahmedabad on the same lines as in Bombay, and it was requested that that they should erect an observation tower equipped with a telephone, supply a smoke observer to work under the control of the resident

inspector and report cases of excessive smoke emission. Bombay SNC, *Annual Report*, 1931–32, 1937–38.

137. Bombay SNC, *Annual Report*, 1935–36. The report also noted that references forwarded by the municipality were on the decline.
138. Bombay SNC, *Annual Report*, 1940–41. Meanwhile, smaller factories continued to grow in numbers, with several silk-weaving concerns that employed small boilers coming up in 1939–40.
139. Bombay SNC, *Annual Report*, 1928–29. The report mentioned that the proportion of carbonic acid gas in a normal atmosphere should be 40 parts in 1,00,000 parts of air.
140. Bombay SNC, *Annual Report*, 1934–35. This was the first such study since the Act had come into effect in 1913, the commission noted. However, there might have been some more preliminary studies done elsewhere. C. L. Dunn and D. D. Pandya, in their study of hygiene and public health in India, had reported in 1925 that owing to the long, dry periods without rain, the air in many parts of India was polluted by large amounts of organic and mineral matter in the form of dust, with as many as 1,80,000 dust particles per cc having been found in the air of Cawnpore in the dry time of the year. An analysis of this dust showed that it consisted mostly of particles of organic matter. See C. L. Dunn and D. D. Pandya, *Indian Hygiene and Public Health*, Calcutta: Butterworth & Co., 1925, p. 44.
141. By comparison, 7,500 particles per cc were reported for Stoke, 11,000 particles for Hull and 30,000 particles for Sheffield in England. On the different parameters used to define air pollution—soot fall per square mile or particles per cc—and their limits, see Gugliotta, 'Smoky City', pp. 102–03. By the early 1930s, she suggests, the smoke inspector of Pittsburgh began to use the criteria of 'hygienically pure air'—air pure enough to be healthful but not absolutely particulate free. This, however, required sophisticated exposure studies, which were not yet possible.
142. However, in one view, this may not have been an entirely accurate deduction as in the mill area tremendous amount of cotton dust was also discharged, which too possibly contributed to the number of particles.
143. 'Pollution of Air in Bombay', *ToI*, 4 Sep. 1935.
144. Bombay SNC, *Annual Report*, 1937–38. These measurements seem to have been taken through the initiative of Dr Hariprasad Desai,

chairman of the sanitary committee of the Ahmedabad municipality, and the conclusion they arrived at was that pollution of air by mill smoke in the mornings in Ahmedabad was 4.5 times more than in Bombay. In time, bacteriological analysis of street dust was also done (along with that of soil and manure) to determine the nature of pathogenic anaerobes contained in them. 'Smoke Nuisance in Ahmedabad', *ToI*, 22 Mar. 1938; BMC, *Annual Report,* 1941–42, p. 340. However, Dr Desai's long-standing efforts in this regard were seen by some as being a cry in the wilderness, failing to find adequate support within the municipality, with smoke continuing to have adverse health impacts. 'N. Daskroi Local Board Scandal', *BC,* 5 Apr. 1938.

145. 'Smoke Nuisance in the City', *ToI*, 14 Nov. 1939.
146. Bombay SNC, *Annual Report*, 1922–23.
147. Declining smoke was the general narrative told through the years. In such cases where the smoke increased over the previous year, commissions made efforts to point to the proximate causes (usually beyond their control, for instance, greater use of inferior coal given relative fuel prices) that may have caused the increase.
148. Bengal SNC, *Annual Report*, 1920.
149. Bengal SNC, *Annual Report*, 1921, 1922.
150. Bengal SNC, *Annual Report*, 1922. From about 200 minutes of black smoke in an hour, it went down to 40 minutes in an hour by 1925, representing a daily reduction of 5,000 minutes. Bengal SNC, *Annual Report*, 1925. However, the nuisance had not quite ceased to be a matter of concern. Thus, writing in the *Statesman* in the context of the building of a new cantilever bridge over the Hooghly, 'a Special Correspondent' noted: 'It is not for a moment suggested that there is anything intrinsically wrong in permitting ships close into a city, although the smoke and noise nuisances are matters which no civic authorities can afford to disregard.' 'Greater Calcutta', *The Statesman*, 5 May 1926. Making a comparative assessment, the *Times of India* had earlier written that early improvements in Calcutta notwithstanding, the city was murkier and more smoky than Bombay, going on to add, 'this is evidently one of those things that they still do better on this side'. 'Current Topics', *ToI*, 17 Feb. 1921.
151. Bengal SNC, *Annual Report*, 1927. Emphasis mine.
152. Bengal SNC, *Annual Report*, 1930.

153. Bengal SNC, *Annual Report*, 1932 and 1933. Even in the USA, Uekötter writes, notwithstanding greater recourse to the economic efficiency argument, the coercive power of local ordinances remained indispensable. Any relaxation in this could lead to quick slide back to previous smoky conditions. Uekötter, *Age of Smoke*, pp. 73–74.
154. Bengal SNC, *Annual Report*, 1934.
155. Bengal SNC, *Annual Report*, 1938. Of the 34 prosecutions that were launched, two were withdrawn. In one case, the owners shifted, while in the other, the requirements of the Act were complied with in the meantime.
156. Thus in 1939, conviction was secured in 30 of the 50 prosecutions launched, one was acquitted, one was withdrawn for want of proper evidence and one was warned. Seventeen cases were withdrawn because the requirements of the Act were complied with in the meantime. In 1940, conviction was secured in 35 of the 41 cases in addition to five cases being withdrawn because due compliance had been done. With the onset of the Second World War, however, the number of prosecutions showed a sharp downward trend again, with 11 prosecutions being undertaken in 1943 and 15 the next year. Bengal SNC, *Annual Report*, 1943, 1944.
157. Bengal SNC, *Annual Report*, 1939, 1940.
158. Bengal SNC, *Annual Report*, 1934.
159. Bengal SNC, *Annual Report*, 1931.
160. 'Smoke Nuisance', *ABP*, 12 May 1932. Also see Kabita Ray, *History of Public Health: Colonial Bengal, 1921–1947*, Calcutta: K. P. Bagchi & Co., p. 301.
161. Indeed, so peeved was the *Patrika* by the failure of the commission that it was willing to recommend that the Retrenchment Committee should wind up the commission and thus save the government half a lakh of rupees that it cost to maintain the commission.
162. Reported by the vice-president of the health association of Ward IV. See 'Smoke Nuisance in Calcutta', *ABP*, 12 May 1932.
163. See 'The Smoke Demon of Calcutta', *ABP*, Health Supplement, 5 Feb. 1939. This citation is from the prefatory remark with which the newspaper introduced the article by Lt. Col. A. C. Chatterjee, director of public health, Bengal.
164. Chatterjee, 'The Smoke Demon of Calcutta'.
165. This claim must be taken with a pinch of salt, given that the commission simultaneously noted the difficulties in keeping abreast

with latest developments globally, war conditions standing in the way of 'rapid progress in this direction', as also the fact that average emissions increased from 0.08 minutes in 1941 to 0.12 minutes in 1944. Bengal SNC, *Annual Report*, 1943.

166. Bombay SNC, *Annual Report*, 1931–32.
167. Bombay SNC, *Annual Report*, 1922–23.
168. Bombay SNC, *Annual Report*, 1923–24.
169. Bombay SNC, *Annual Report*, 1925–26.
170. Bombay SNC, *Annual Report*, 1926–27. On the other hand, prosecutions sanctioned against the Commercial Ahmedabad Mills, City of Ahmedabad Spinning and Manufacturing Company Ltd and the Ahmedabad New Cotton Mills Company Ltd had to be withdrawn on representations being made by the respective mills.
171. Bombay SNC, *Annual Report*, 1930–31.
172. Bombay SNC, *Annual Report*, 1931–32.
173. Bombay SNC, *Annual Report*, 1934–35. One of these could have been Jam Mill No. 1 of Lalbaug as reported in *BC*, 9 Jun. 1934.
174. Bombay SNC, *Annual Report*, 1935. We may recollect that one of the chief complaints in the nineteenth century was that frequent warnings were not accompanied by penal action, allowing mills to continue to emit excessive smoke. The mention of seven consecutive offences would suggest something similar now, though the *Bombay Chronicle* felt otherwise, writing that the extent of the cumulative fine suggested that 'smoke is a costly affair, especially when it is dense black'. 'Smoke Nuisance', *BC*, 29 Jul. 1936. Around this time, there was also a dispute over whether owners were personally responsible for excessive smoke or should only the engineer in charge or the stoker be held responsible. 'Responsibility for Mill Smoke Nuisance: Point of Law Decided', *ToI*, 3 Aug. 1934.
175. Bombay SNC, *Annual Report*, various years. And even then, not all prosecutions were successful. See, for instance, Bombay SNC, *Annual Report*, 1939–40, 1945–46. Occasionally, the matter would also reach the courts via the provisions against nuisance in the Indian Penal Code, as, for instance, in April 1937 when Uttamchand and Balkishan of Malegaon Camp initiated proceedings against the owners of the premises adjoining a flour mill, pleading that the walls of their homes shook because of the working of the mill, while the sound and smoke from the engine caused nuisance. See

Bombay High Court, Emperor vs Rameshwar Narayan Agarwal on 2 Nov. 1938, (1939) 41 BOMLR 84.

176. Firemen did not seem to appreciate the value of these certificates, the commission ruefully observed, giving more importance instead to a 'third class' Boiler Act Certificate, though both certificates required about the same amount of intelligence. Bombay SNC, *Annual Report*, 1919–20.
177. Bombay SNC, *Annual Report*, 1933–34.
178. This remark was made by Fakirji Bharucha who was to serve as the President of the Institute of Mechanical Engineers. See, 'Power Costs for Bombay Mills', *ToI*, 11 Feb. 1937.
179. Bombay SNC, *Annual Report*, 1927–28. Next year, the number of ships had increased to 45, and mention was made of the beneficial results obtained through advice regarding fuel use. Depression years, on the other hand, saw a decline in shipping at the docks and the smoke nuisance in the harbour correspondingly became much less. Even when trade recovered by the mid-1930s, smoke at the harbour was reported to have remained within permissible limits, the commission observing that the few offenders who remained were served with prompt notices 'which had the desired effect'. Bombay SNC, *Annual Report*, 1932–33, 1934–35.
180. Though we learn of persistent difficulties in this regard too. Thus in 1933, the Bengal commission observed that owing to the industrial depression and the general demand for economy some of the larger concerns that had been pressed to improve their steam-raising equipment in order to lessen atmospheric pollution failed to improve their plant. See Bengal SNC, *Annual Report*, 1933, 1934. Similarly, in Ahmedabad, the commission observed that while there had been an overall improvement, few of the larger mills still remained an exception. 'It appears that the more prosperous a mill is, the more indifferent its Agents are about smoke emission', they noted. Bombay SNC, *Annual Report*, 1925–26.
181. Bengal SNC, *Annual Report*, 1944.
182. Bombay SNC, *Annual Report*, 1921–22.
183. Bombay SNC, *Annual Report*, 1952–53. Cumulatively, reduction of black smoke per day was reported to have been from 8,000 minutes in 1925 to 2,000 minutes in 1930. Bombay SNC, *Annual Report*, 1931–32.
184. Bombay SNC, *Annual Report*, 1939–40, 1940–41.

185. This, we may recall, had been a constant practice in Indian industrial settings. Speaking specifically of cotton trade, Cecile Matheson wrote that 'discipline is enforced by fines or dismissal and, unofficially, there is a good deal of corporal persuasion in some mills'. Cecile Matheson, *Indian Industry: Yesterday, To-day and To-morrow*, London: OUP, 1930, p. 37.
186. Bengal SNC, *Annual Report*, 1941.

5 Home and the Hearth

The Domestic Smoke Question

No fire-place, you will say; do they eat your raw rice then? To this I have two answers: first, that the kitchen with us, as with our neighbours is not in the house, but in the compound ... And next, that the kitchens, without a single exception, are without fireplaces. These black Indians, who are our cooks, are accustomed to light the fire without caring for the smoke; which escapes where it can, through the windows, or the skylight, or through the holes in the roof.[1]

Calcutta lives among the chimney tops. It's quarter of a million of fireplaces, tens of thousands of shops, eating houses, sweetmeat shops have no chimneys. The smoke and poisonous gases are discharged at the ground level ... and often remain trapped here on account of atmospheric conditions.[2]

Labour, Housing and the New Social

Calcutta and Bombay, as we saw in Chapter 1, had for decades struggled with the housing question and with its sanitary dimensions—fresh, circulating air inside homes and fresh, free-flowing breeze outside on the streets. The years following the First World War saw these issues being rearticulated, alongside greater intervention in issues of labour reform, undertaken in the background of overcrowding and congestion on the one hand, and greater organisation of labour through unions and strikes across industries on the other.[3] Demands of labour, management of space and the quality of urban life thus came together to forge a new social question around housing. Managing urban air was reintroduced as an important sanitary agenda, with Government of India proposing to fix standards of light and ventilation to ensure sufficient outside air space for every room and 'then to ensure that the standard shall be enforced and no building permitted in contravention thereof'.[4] Providing decent homes and educational facilities for the children of the working classes as much as negotiating 'fair' wages and creating more conducive working atmosphere featured together in strategies of mill owners, governments and, increasingly, labour unions, both in Bombay and in Calcutta.[5] These connections had been flagged earlier too, especially in the post-plague years, when industrialists had sporadically articulated the need for decent housing for workers to ensure that they did not flee the city.[6] However, the translation of these sentiments into actual construction rarely matched. Some chawls were indeed constructed in Bombay, but many within the community of mill owners were convinced that workers' housing, properly speaking, was a task to be carried out by the Improvement Trust and the government. There was no doubt that the health of the employees' suffered considerably owing to their unhealthy surroundings, Fazulbhoy Currimbhoy Ibrahim had argued, but it was not for the mill owners to build the chawls, given the low returns on such construction. In his testimony to the Indian Factory Labour Commission (1908) he therefore submitted that industrialists 'could not become philanthropists at the expense of their shareholders'.[7] In Calcutta, on the one hand, the chairman of the Indian Jute Mills Association was in favour of building pucca brick houses for workers, such buildings being considered more sanitary, while on the other, the manager of Hastings Jute Mill was of the view that no law should be framed that would compel mill owners to provide housing to their labour.[8] By the first decade of the twentieth century, Bombay's working class population thus continued to live in overcrowded, ill-ventilated chawls with poor sanitation and water supply, and by the end of the war years, 'house famine', affecting particularly the working classes, was seen as Bombay's most acute social problem. In Calcutta too, where some housing was provided by mill owners who found cheap land on the outskirts of the city and hoped to benefit from a stable working population resident in buildings attached to the mills, the majority continued to live in private bustees, in the city and on its margins. Better housing for workers, the Industrial Commission thus urged, was 'a most urgent necessity', especially in the large congested cities, for both moral and economic reasons.[9]

The immediate post-war years thus saw a renewed engagement with the housing and labour question, prompted as much by local economic boom as by international conventions agreed upon at the first International Labour Conference in 1919 and the urging of the liberal press in England that 'interest in humanity need not stop with the shores of England'.[10] In part, the response was in terms of design, to determine what kind of housing would be best suited for the working classes.[11] In part it was institutional, with the Improvement Trusts continuing to be active both in Bombay and in Calcutta, along with new governmental agencies, such as the Bombay Development Department (BDD), justifying their intervention as an aspect of benevolent rule and in the interests of the most vulnerable.[12] Notwithstanding this rhetoric, the Improvement Trusts, which had been much criticised in the pre-war years for focusing on thoroughfares and wide streets rather than on sanitary housing, continued to perform as badly post war, with wages and rents remaining mismatched. The Development Directorate too soon began to make the excuse that its initial target of building 50,000 one-room tenements for the working classes (over eight years) had been wildly optimistic and changing conditions demanded that these be drastically revised downwards.[13] Indeed, they doubted very much if such a shortfall of working-class housing existed at all![14] Working-class housing in Bombay thus increased by less than 5 per cent between 1921 and 1931, in no way commensurate with the increase in population. And even there, the three mill districts saw a particular intensification, even as population declined in the older native settlement areas.[15] The municipality pointed to the empty homes and reduced rents, but the *Bombay Chronicle* rhetorically asked 'can the worker afford even the reduced rent that is demanded?'[16] The answer, based on all available evidence, was in the negative. Calcutta similarly had little to offer by way of the activities of the Trust regarding the housing of the urban poor.

On the one hand, the Calcutta Improvement Trust had much to write in its own favour—of having made Calcutta healthier and more pleasant, even receiving the endorsement of prominent nationalists such as Surendranath Banerjea and Bipin Chandra Pal. But on the other hand, there was little denying that the bustees had been hard hit in this process of improvement—by 1927, half a square mile of property had been demolished and no less than 50,000 people displaced.[17] Housing measures taken up for the poor had abysmally failed, and the utilisation of improved lands for bustee purposes ruled out as being uneconomic. By the 1940s, the Calcutta Imrovement Trust itself conceded that 'Calcutta has been up to the present a city without a housing policy ... there is at present no organization for the re-housing of Calcutta as a whole'.[18]

And so it would be that chawls and the bustees, much discussed in terms of the responsibility of mill owners or government and as regards general sanitation, also became sites for considering the question of smoke. Officials routinely suggested that the difficulty of maintaining chawls in a sanitary condition was owing to the habits of their residents—throwing *kutchra* or refuse from their windows instead of letting air in.[19] On other occasions, more sympathetic land managers noted that the lack of sanitation was not on account of any fault of the residents but because of the imperfect working of the chimneys in one-roomed tenements, combined with inadequate ventilation of latticed windows, which ensued darkened rooms, with the smoke and soot remaining gathered indoors.[20] Labour representatives and bhadralok activists additionally put the blame on the 'industrial system' that condemned workers to life in slums, working long hours in a closed and stuffy atmosphere and whose wages rarely matched the rents demanded.[21] Debates on smoke—its impact on residents, especially women and children—intersected with this debate on culture and political economy regarding sanitation generally, portraying residents as victims, either of industrial smoke that entered their homes or smoke produced in kitchens that had little or no room to escape. On the other hand, there was also the issue of domestic smoke that potentially harmed the city at large.

Industrial smoke could well ebb with trade cycles or with labour strikes that resulted in periodic shutdown of mills, but domestic smoke was a constant, which is what made it more pervasive and yet difficult to contain, in India as in England:

> If you go to a housetop in a busy quarter of a populous city, like Bombay, Calcutta, Cawnpore or Delhi, especially in the evening when the air is still, you will notice the dark trails of smoke and copper colored haze in which the city is enveloped ... The phenomenon of smoke filling all the houses in the evening in winter, causing smoking sensation and smart in the eyes is due to the waste products of the lacs of choollahs filling the air and kept imprisoned just above the ground due to absence of breeze and heaviness of the cold, dank, air![22]

In this latter context, the concern would be with urban health in general and the solution sought more among the domestic arrangements of the middle class than in the quality of life in bustees and the chawls.

It is these narratives of smoke inside homes and outside in the city that I attend to in this chapter, suggesting that while domestic smoke as a harm for workers who lived in them was more pronounced in Bombay (where more investigations into labouring conditions continued to be conducted, especially following the first International Labour Conference in Washington in 1919), the question of domestic smoke as a harm to the city was far more evident in the concerns of the Bengal Smoke Nuisances Commission. Bombay, as we suggested in Chapter 4 too, remained obsessed with the small-scale industries instead, which, to reiterate, were often also located in dense residential neighbourhoods and were, on that count, productive of much nuisance. Both cities, however, would offer a common recipe for the ultimate triumph over domestic smoke—the greater use of gas and electricity, though the extent of its actual adoption and impact, across classes and in different localities, would leave much to be desired. Domestic smoke, in other words, like industrial smoke, was at once a generalised issue and also marked by class, gender and spatial differences.

Smoke Inside Homes

Smoke in domestic settings, as we saw in Chapter 1, featured prominently in accounts of vitiated air. And in this, the home was marked rather differently than the factory—if in the latter, there was dust inside and smoke outside, in the former, both dust and smoke could be found together inside. Indoor air was thus fouled on more than one count, being simultaneously about sanitation, ventilation and pollution, prompting continual comparisons between chawls and mill spaces through the twentieth century. Comparative assessments of mills and housing had not entirely been missing in earlier enquiries either, though done mostly from the perspective of ventilation and sanitation. The commission appointed to consider the working of factories in the Bombay presidency in 1885 offers numerous witness accounts of mill agents whose general refrain was of the chawls faring worse than factories.[23] Most members with industrial interests were also convinced that diseases such as colic, constipation, diarrhoea, fevers and muscular pain were not on account of factory work but because of conditions outside. However, there were two interesting exceptions. K. N. Seervai, who had served as the inspector of factories for a brief term between the tenures of W. O. Meade-King and James Jones, was convinced that in actual working terms, 'when all the doors and windows of a room are closed', there was less ventilation in a mill than in the work people's chawls or dwelling houses.[24] Narayan Meghaji Lokhande, editor of *Din Bandhu* newspaper and chairman of the Mill Hand Association, was similarly of the view that while in theory, there was possibly better provision for ventilation in a mill than in the chawls, in practice, it was established that windows in the mills were used more for light than for ventilation.[25] And even if one were to assume that people were as closely confined as in the mills (which they were not), even then the temperature inside the mills would be much higher and the air more 'vitiated'.[26] Mill owners and their representatives were not to be persuaded though, continuing to argue before various labour enquiry committees instituted in the first decades of the twentieth century that workers fared much worse in their own homes than in mills. The point was put most forthrightly in a letter submitted by A. K. Leslie of Messrs Greaves Cotton and Company:

> Atmosphere of cotton mills is subject to much misrepresentation. True the atmosphere of mills is not as pure as that of Malabar Hills but most mills are well ventilated, and infinitely better than the atmosphere prevailing in the homes of the operatives and is no worse than the atmosphere most common to Bombay. The whole question of atmosphere must be one of comparison, and I am strongly of the belief that the atmosphere of our cotton mills is much more free from harmful constituents than the atmosphere of the rooms in which our mill-hands live. I do not advance this as any reason why we should not strive to improve the hygienic conditions of our mills; but I do consider that it is time that the question of the atmosphere of our cotton mills should not be regarded from the sentimental side, or on a beatific basis only.[27]

This may have been an exaggerated stance perhaps, as other studies found conditions inside factories much wanting. More importantly, even if the mills were to be granted 'better' conditions inside, could they still be implicated in the more adverse conditions outside? The question of smoke would suggest that such links were indeed posited by contemporary observers.

Workers' homes, the Census of Bombay noted in 1901, were not only blighted by the lack of sewerage, but also by the presence of thick black smoke from the coal-powered cotton mills, railway locomotives and government mint. Lingering dark clouds, the Census wrote, brought about 'a marked change ... over localities such as Byculla, Parel, Tardeo, Tarwadi and even remote Sewri'.[28] The *Times of India* similarly drew attention to the condition of the chawls, observing that at least some of the adverse atmospheric conditions in which workers lived could be attributed to the functioning of the very same mills in which they worked:

> To realize how the vitality is being crushed out of the workers in the Bombay mills ... you must sally forth in the darkness, and watch them swarming along the roads,

> dim, swathed, melancholy figures … Many of the Bombay mills are situated in the least healthy parts of the island. In the first hours of daylight, and after the sun has set, mephitic exhalations arise from the adjacent swamps and pools … The atmosphere of the district is unwholesome and tainted by effluvia, to which hundreds of belching chimneys add their quota of dense volumes of foul smoke. The external precincts of most of the mills are evil smelling and repellant.[29]

Studies of labour in the post-war period, both official and based on independent research, continued to draw attention to this feature. Among the most influential of these reports was that of Dr Florence Barnes, submitted to the Bombay Legislative Council in 1922, and of Dr Dagmar Curjel titled *Women's Labour in Bengal Industries*, conducted the same year.[30] Housing in Bombay, especially for the working classes, Dr Barnes observed, was extremely unsatisfactory. In one such room, measuring 15 feet by 12 feet, she found six families living, each with a separate oven for itself, a small brick erection made with three bricks placed at right angles to each other and without chimneys. Privacy for pregnant women was ensured by screening off of a small space with the use of old rags and sacks, leading her to observe that the atmosphere of the room at night, 'filled with the smoke from 6 ovens and other impurities', would certainly handicap both the woman and the infant before and after delivery. Conditions were even worse in the basement, where 'daylight with difficulty penetrated, sunlight never'. The atmosphere of factories could well be laden with fibre, but even that was better than the homes of workers, 'where every ventilation space was packed to avoid the ingress of fresh air, and where, for fourteen hours of the twenty-four, the family inhaled an atmosphere laden with smoke and other impurities'.[31] Dr Curjel, by contrast, found little to report on the cooking arrangement inside homes, confining her remarks to the more traditional sanitary concerns, the 'considerable fluff and dust' in the older jute and cotton mills and the lack of circulation of air inside, which had a bearing on both health and productivity.[32] The housing provided by mill owners, especially for migrant labourers, she reported, was defective, with rooms placed in a line, back-to-back, and with no proper arrangements for ventilation, this being specially worse in double-storeyed houses. Some mills did provide open courtyards, especially for higher paid workers, but sub-letting was common and Curjel made a special mention of the extent to which this robbed women of their privacy. Houses located at a distance from the mills, on more open tracts, seemed to have fared better in this regard, both with respect to privacy and as regards ventilation, though even about these, she wrote, the surrounding areas were often insanitary.[33] Curjel made no mention of smoke though, leaving it to other observers to conjoin homes and smoke therein, especially with regard to women's household burdens and their health. Urban women, L. S. S. O'Malley informed, were more prone to suffer from the impairment of sight than their rural counterparts, for the reason that while village homesteads were well ventilated, in Calcutta, women had to spend their time mostly indoors, much of it in 'small dark rooms filled with the acrid smoke of cow dung fires, at which they cook their food'.[34] O'Malley, in turn, drew upon the observations of Dr H. M. Crake, who wrote:

> I must confess I am astonished at the average kitchen. It is in a large number of houses, a gloomy, stuffy den, full of acrid smoke, and yet the ladies of the house have to spend hours in these very unpleasant surroundings. The entire absence of chimneys results in an atmosphere which is almost unbearable when cooking is going on in a particularly ill-ventilated kitchen.[35]

The injurious effect of dust, both in Calcutta and in Bombay, was also being gradually recognised. Employers, Janet Kelman wrote, were aware of the issue but remained content with the observation that escaping dust in India was an impossibility, within or outside the factory.[36] Indeed, given these dusty conditions, it may well be asked 'why trouble about a little fluff?'[37] To which she proceeded to answer:

> Fluff laden air ... has special dangers for health, and the advantages to be gained by the introduction of exhaust

> fans and ducts on the machines from which the greatest amount of fluff emanates can scarcely be over-estimated… Nothing but the experience of hours spent in the stifling atmosphere of the worse ventilated mills can make it possible to realise the exhaustion and lassitude that result. Yet, in these conditions men and women labor through long hours, and little children doze in their hammocks or gaze on their world from half-hidden corners.[38]

And as with Dr Barnes, in her account too, the conditions of homes and smoke inside homes figured prominently, especially as they affected women. Describing the chawls of north Bombay, she wrote:

> It is very difficult to give any clear idea of what a group of very bad chawls are like … As the eye glances upwards, it may see a wall variegated by rags, boxes and matting. These indicate the stuffed up windows of chawls that have inner tunnel like passages from which the occupants enter their houses … The rooms behind these stuffed up windows and draped verandahs are dark … The heaviest burden of such conditions falls on the woman [who] … exhausted from her long day's work, huddles down to sleep in the dark, unventilated room along with women lodgers and relations and their children and her own … In the stuffy atmosphere of the room, a fire has to be lit, and if a lodger happens to be of a different *jat* [caste], a second fire may add its quota of smoke in another corner. The pungent smoke from cow dung cakes or wood finds its way out as best as it may. It is after such a night and morning that a woman worker finds her way to her day's employment at the mills. It is not to be wondered at that the infant mortality in Bombay is high.[39]

G. M. Broughton, in a doctoral thesis written for the University of London around the same time as Kelman was carrying out her research, was much briefer in his observations on the housing question, though he too, like Dr Barnes and Kelman made a special mention of the labour that fell on the woman resident. An Indian woman, he wrote, took great pride in keeping her utensils clean, these being one of the few possessions the family had. However, this was no longer possible in the insanitary chawls of Bombay or the crowded bustees of the Calcutta suburbs. 'All the cooking, with its attendant smoke soon blackens everything', he observed, 'for chimneys and fireplaces are unknown and even the clothes that are hung up to screen the verandahs very soon become dirty.'[40] Thus as has been observed in other contexts too, while the health risks associated with coal smoke were distributed unequally between the classes, the human labour at home and work impacts of polluted air were divided unequally along gender lines, time consuming and strenuous house cleaning and the washing of clothes pushing legions of women into a never-ending round of drudgery.[41]

More studies followed. In 1925, the Indian Research Fund Association, Shimla, awarded a grant to Margaret Balfour and Shakuntla Talpade to follow up on the earlier study of Dr Barnes on the maternity conditions of mill workers and other classes in Bombay. Once again the focus was on facilities inside mills (for instance, availability of crèches) and conditions outside, in the chawls. The description offered was not very different than what had been offered earlier, and even though the extent of overcrowding pointed out by Dr Barnes may have been extreme, Balfour and Talpade also estimated that the average space at night per head was 240 cubic feet.[42] The rooms themselves, were reported to be usually 12 by 15 feet, organised in rows, with a central passage in between and occupied by one or more family. They usually had a window opening on the street, but the only other air space, the door, opened into the passage, not into the outer air. The one window also did not open directly to the air but into a verandah where the family cooking was done, so that the air that entered the room was always contaminated with smoke. In time, others would also share their view but the proposition that it was the habits of the working people that led to ill-health was never too behind either. A more detailed study of labour and housing in Bombay, carried out over three years of field work by Burnett-Hurst of Allahabad University, for instance, reported at length on the chawls, and on sheds of corrugated iron and zavlis (constructed using dry leaves) in which the workers lived, the first housing used by nearly 70 per cent of the entire workforce of the city, the second kind occupied by Mahars, Chamars and Dheds (members of the Depressed Classes) and the last by Ghati carters.[43] The chawls, he pointed out, were of three kinds, with ample ventilation and light only available in the most recent, modern type of chawls built by the Improvement Trust, municipality and other public bodies, for their own employees or for the working classes generally. And given this favourable perception of housing provided by public authorities, the study naturally turned to the cultural habits of the workers themselves, as a reminder of the environment that they had created.[44] The external appearance of the chawl may be neat and attractive, Burnett-Hurst wrote, but an examination of the interior startled one. All sense of cleanliness appeared to be absent. Spitting of *paan* and betel nut juice and other nuisances were committed everywhere, especially on the staircases, and in the passages and corridors. These places also served as the chief repositories for the sweepings of the rooms. Goats, fowls and other animals belonging to the tenants were also often to be found in the corridors. It is in these surroundings that babies crawled, children played and mothers nursed their infants! The interiors of the rooms were little better than the rest of the building, he further observed. Windows and sometimes verandas were closed with rags, gunny cloth, clothing hung out to dry, etc. Where woodwork and glass ventilators were provided (for example, in the Improvement Trust Soparibag chawls), they were frequently blocked with firewood. And even where verandahs were fitted with cooking places, tenants preferred to cook in their living room, especially on a windy day. It was not surprising, then, that the walls and ceiling were blackened, there being little escape for the smoke. Not surprising either that disease abounded, death rates were high and infant mortality excessive. And for all this, it would seem, workers themselves contributed 'in no small measure'. 'It is no exaggeration to say', Burnett-Hurst concluded, 'that the masses are utterly unacquainted with even elementary ideas of hygiene and sanitation, and little improvement can take place until they have been educated to a different standard of living.'[45] In this, he did no more than mirror the language of officials in charge of developing Bombay who masked their own failures with comments that frankly betrayed their class and caste prejudices![46]

Not all were as unsympathetic towards working-class cultures, however, using the same 'facts' to suggest more the difficulties faced by workers rather than their habits in creating insanitary microenvironments. Bad enough as the question of housing was in descriptions provided by Burnett-Hurst and later by the Royal Commission on Labour, wrote G. R. Pradhan, the lot of the Depressed Classes was even worse.[47] Their houses presented an even more dismal appearance. Verandahs, if present, were converted into a bedroom by fixing up gunny bags all around it. Protection from the sun and the rain was similarly obtained using worn-out pieces of gunny or even of rags on the windows and doors, which became extremely dirty on account of the use of sawdust, dung balls, waste paper, inferior wood and more rarely coal as fuel. Naturally the room was full of smoke—walls, roofs, pillars and, in fact, everything in the room became dark, and one could only imagine how much the residents suffered from the smoke nuisance. This was not all. At night, kerosene-oil tin pots were used as lamps, and with no glass to serve as a chimney, their users suffered from the injurious effects of carbonic acid gas.[48] If there were two or more than two families living in one room, there were as many *chula*s as there were families, producing heat and smoke. And in these heated and smoke-filled rooms, little children slept, school-going children studied, and the sick and the tired took rest.[49] In other chawls, food was cooked in the small, narrow and dark passage outside the room. In one way, this was good, Pradhan wrote, as the room was free from smoke and heat, but cooking outside the room presented its own gloomy spectacle, intensifying the darkness of the passages when they were full of smoke. In this darkness, 'many *chulas* twinkled' and 'the long passage between the two rows of rooms appeared to be some devil's den and not a residence of human beings'.[50]

Ventilation and smoke in Calcutta bustees remained equally problematic—housing in the six mills investigated in 1930, for instance, suggested overcrowding and inadequate ventilation in quarters provided by the companies and, even more so, in private bustees adjoining the mill areas. Smoke was not directly mentioned, but the fact that cooking was usually done in the one room used as living quarters would suggest that the situation would not have been very different from Bombay.[51] The use of particular fuels would not have helped matters either, the same report suggesting that though coal was cheaper than firewood, 'some sections, as the Madrassis, owing to religious scruples universally use firewood', while kerosene oil was used for lighting.[52] As before, insanitation, smoke and disease went together, the extreme cases of which could even result in death.[53] But what obtained in descriptive accounts still struggled to find a place in the medical reports of Calcutta and Bombay, despite the fact that a large number of deaths from respiratory causes had always featured prominently in the health statistics of these cities. In Calcutta, such deaths were often reported to be on account of (i) poverty, under-feeding, the struggle for existence under adverse conditions, etc., (ii) ignorance and carelessness resulting in sputum loaded with the bacilli being expectorated all over the place, (iii) bad housing, a dark, damp, ill-ventilated hut in an open suburb being just as deadly as a badly lighted and ill-ventilated room in the heart of the city, and (iv) the purdah system.[54] Other causes included, more generally, poorly drained suburbs, the poverty of bustee dwellers, overcrowding and lack of fresh air. In Bombay, similarly, high death rates even in non-epidemic periods were usually associated with 'the insufficiency of the sewers, the continuance of the basket privy system of conservancy, the density of houses and persons per acre and the overcrowding of persons in rooms and tenements'.[55] Missing from both was smoke as a factor of disease, a link that was made by health officers only by the mid-1930s:

> Suburban wards, particularly those which are ill drained and are liable to flooding, show high death rates from

> respiratory diseases year by year. The poor bustee dweller, ill clad, ill fed, living in damp huts, is particularly prone to these diseases. Overcrowding and want of fresh air are also important factors for the development of these diseases. Last though not the least in importance is the smoke nuisance which is a prominent factor for carrying away many lives from respiratory diseases.[56]

This had been a cry within the Smoke Nuisances Commissions for more than a decade now, though more at the level of the city as a whole than at the level of the individual household in the chawl or the bustee. 'One fifth of the total death rate in Calcutta is due to diseases of the breathing organs. Medical authorities suggest link with smoke', the Bengal commission had observed in 1926.[57] And the cause for this continued to be attributed to the peculiarity of housing design in India, particularly the absence of chimneys. In Europe, the commission pointed out, such smoke was discharged at an altitude of 50 to 100 feet, or even higher, depending on the height of the buildings, each of which had suitable chimneys, and, ordinarily, there was a zone of comparatively pure air near the ground for breathing purposes. But in Calcutta, all such smoke was discharged at ground level, the community living in and having to breathe the poisonous sulphurous air. This was as true of the middle-class home as that of the urban poor, indeed of European homes in India too.[58] Remarking on the inherent conservatism of the Indian cook, an article in the *Statesman* opined: 'Give him an earthenware, *chula* on top, earthen dishes, and a cookhouse full of acrid smoke, and the *bawarchi* is as happy as the proverbial king.'[59] This, we are further told, was in the case of European houses. In Indian homes, cooking arrangements, 'as primitive as in the days of Clive', left even more to be desired, there often being no separate cookhouse and no escape for the smoke. The consequence: 'The residents in them have ... suffered from the smoke nuisance in the most virulent form for years; for the smoke from the *chula* cannot be confined to the kitchen; it diffuses throughout the house and lingers for hours after each meal.' Indians, as much as Europeans, worried on this count. In time, therefore, domestic smoke would be addressed by the Smoke Nuisances Commissions too at some length, especially in Calcutta, but without quite going into the structural features of individual households in which living rooms doubled up as cooking spaces in the poorer households, and where children and women were routinely exposed to the hazards of indoor smoke. Also, while in the instance of industrial smoke, much would be by way of transfer of European expertise to India, the special conditions of Calcutta's domestic smoke, 'having due regard to economic factors', would require 'special measures', if mortality was to be reduced and the public health improved.[60] This, however, was never to be, at least for the poorer households. Instead attention shifted away from the individual home to the city at large.

Domestic Smoke and the Urban Setting

Smoke that escaped from homes onto streets had also attracted attention for long. However, for much of the nineteenth century, the issue had remained on the backburner, leaving it to Grover to single out domestic smoke as an important contributor to the nuisance in the city:

> In the case of native huts the fuel used burns with a peculiar pungent odour. It is for the most part burnt within the dwelling and the smoke from it rises to the roof through which it percolates. The larger particles of soot are thereby deposited on the under-surface of the roof, and it would appear from this that, apart from objectionable odour, there can be but little deposit of soot. Such smoke being situated near the ground level affects more particularly the inhabitants in the immediate vicinity of the dwellings.[61]

Grover's preferred solution was the isolation of areas responsible for the emission of such smoke, a suggestion that seems to have found little favour. The British Indian Association countered that domestic fireplaces in native huts contributed 'but an inconsiderable part of the aggregate nuisance', and at any rate, the remedy suggested was decidedly worse than the evil. The question of the isolation of native huts, they opined, was intimately connected with the larger question of the housing of the poor and could not be so lightly or easily dealt with or disposed of. The Bengal National Chamber of Commerce further argued that the pungent odour felt in the smoke coming out from the domestic fireplace of poorer classes of Indians was due to cow dung cake, which was not only the cheapest fuel in the town but also far from being poisonous, especially if one compared it to the injuries of coal. The government, it would seem, was not quite persuaded either, excluding domestic smoke from the ambit of the law enacted in Calcutta in 1905 on the argument that isolating native huts would be entirely impracticable. So too in Bombay, at the time of the making of the 1912 legislation, where it was observed that

> here we are not required to consider the smoke which arises from domestic users, which greatly complicates the smoke evil in western countries. Although this is a matter which will probably need attention in the near future, the smoke produced from private houses is now an inappreciable factor. The nuisance arises from the manufacturing industries, which means cotton industry.[62]

However, the view persisted, with Nicholson, the first interim smoke inspector, reiterating that the contribution of native huts to the smoke nuisance was not insignificant. There were fires used for domestic and trade purposes in which all kinds of combustible matter were burnt, he noted, observing again that there was no adequate arrangement of chimneys for carrying away the sickly smoke. The only way to abate these nuisances, as far as he was concerned, was to put out the fires, a remedy even worse than that suggested by Grover! Domestic smoke also began to feature in the reports of the Smoke Nuisances Commission in Calcutta from its very inception, albeit with some degree of regret: 'much of the smoke nuisance is from domestic fires, which it is not possible to deal with under the Act'.[63] Or yet again:

> the smoke from domestic fire continues to be a considerable nuisance, especially during the cold weather season, and

> this is an evil which cannot be remedied under the Act. It is aggravated by the fact that such smoke is discharged almost at ground level, as compared with the discharges from chimneys in European cities.[64]

Newspapers equally made this an issue, with the *Englishman* writing of the 'miscellaneous fuels' that were burnt in Calcutta's bustees, with their 'primitive domestic fireplaces', the smoke from which rendered the air pungent and unbearable.[65]

Domestic fires were not alone either. Within a decade of the Act having been passed, concerns were equally expressed about other sources of ground-level smoke, the commission noting the adverse impacts of open fires, especially those in which raw coal was used as a fuel and of the 8,000-odd trade furnaces that produced smoke of colours other than grey or black, keeping them out of the ambit of law as currently enacted.[66] The smoke from *bhunja* furnaces (used in street vending, particularly for roasting rice flakes) was also the subject of much complaint, and the issue here was even more complex. On the one hand, it was evident that these furnaces used light refuse as a fuel, not an undesirable thing in itself, as it obviated the need for the use of coal or coke in these furnaces. However, this also implied that surrounding houses got covered with soot and smuts, and as most of this trade was in the poorer and more crowded parts of Calcutta, the nuisance caused was considerable. On the other hand, the commission observed, these places were owned and run by comparatively poor people, and the food cooked by them was consumed by the people residing in these neighbourhoods, making their presence a necessity in the localities they worked in. Any severe measure which could result in forcing them to move elsewhere was consequently inadvisable. Prosecutions were sometimes unavoidable, but an arrangement of furnace fitted with soot arresters designed in the commission's office was perhaps a better solution. Several owners were thus prevailed upon to amalgamate to reduce capital cost per head and adopt the improved furnace.[67] The reports, it would seem, were encouraging too, with the complainants writing that 'since we had the occasion to complain we have noticed a considerable reduction of smoke and take this opportunity of thanking you for the trouble you have taken in this matter', and the commission happy that 'in any case, the refuse had to be burned and it is better that it should be used in properly designed furnaces'.[68]

Throughout Calcutta and its suburbs there are many thousands of furnaces that emit smoke, because the owners are ignorant of the means by which it can be prevented. These furnaces are generally used by food-suppliers who cater for the people in their locality. The above photograph shows a furnace smoking in the usual manner and one from which there is no smoke. In both furnaces, coke obtained from the Municipal Depot is being used, and both were lighted at the same time. The one on the right shows the gases or smoke distilled and lost when lighting up or replenishing the fire owing to an insufficient supply of air ; in the other the gases are consumed and utilized through supplying the necessary air by means of a hand-punkha.

The advantages of the smokeless fire are :—

(1) Economy—no waste.
(2) Better food—no smoky taste.
(3) Cleanliness—no damage to health and property.

Source: Bengal Smoke Nuisances Commission, *Annual Report*, 1918–19, Appendix 2.

Figure 5.1: Smoke on the Street: Calcutta

Bombay, by contrast, would have little to say about these trades for the longest time, though it would seem that independent reports did associate local hawking of goods with smoke and other nuisance.

There was also the issue of smoke from coke made in the open. As we saw in Chapter 3, the law regarding this had been amended in 1916, but there was little stopping its continual use. The Bengal commission noted in 1920:

> The scarcity and high price of coal and coke has caused a violent recrudescence in the making of coke from raw coal in open heaps on the ground. This practice is carried on in the neighbourhood where coal is handled, usually at the docks, railways coal depots etc. by a floating population

> of causal labor. The coal is stolen and the coke is sold and owing to the present high rates the business carries big profits.[69]

In the years that followed, several prosecutions were initiated and convictions obtained as a measure to remedy this situation. The police was authorised to undertake proceedings against the actual coke makers in 52 cases in Calcutta in 1921, in 40 of which the commission surveyed, measured and traced the boundaries of the ground upon which the coke was made. It also located the owners, to whom 48 offences were reported, with prosecutions being instituted regarding the remaining four. The next year, 17 warnings were issued for illicit coke-making, and the police authorised to take action in eight cases and the landowners in two cases. Three warnings were given to owners for the making of illicit coke on their premises the year after and the police authorised to take action

Source: *The Times of India*, 17 Mar. 1937: 'The lovely golden sands of the Chowpatty Beach, Bombay, are marred by the hordes of hawkers who congregate there every day. At night the smoke from Naptha flares contaminate the air and in the morning the sands are strewn with litter.'

Figure 5.2: Smoke at the Beach: Bombay

in 24 cases against actual coke makers. In 1924, police action was again authorised in 12 cases.[70] Yet, the commission observed, the progress made was not that which the public reasonably expected, with complaints still received regarding 'the destructive smoke that envelope certain localities like impenetrable fog'.[71] In brief, there were quarter million domestic furnaces and, additionally, thousands of sweetmeat shops, bhunja shops, bakeries and other furnaces in which raw coal was used. Their impact was now to be 'anxiously considered by the commissioners', so as to 'at least make a beginning on the domestic and kindred low lying smoke, which is so deadly to the public health of Calcutta'.[72]

And so by the end of the 1920s, as the commissions gained confidence in their ability to abate industrial smoke, they directed their attention to the low-lying smoke discharged at the ground level from the numerous domestic fires. The former could be more easily seen and blamed, they pointed out, but it was the latter that was 'more deadly' and therefore deserving of greater attention.[73] This was so especially in the suburbs, particularly in the cold weather evenings, when localities were enshrouded for several hours in a low-lying suffocating smoke pall, even on holidays when the mills were not functioning. The figures were amply illustrative in this regard—by 1928, daily discharge from factories and steamers was reported to be about 8,000 minutes. By comparison, domestic smoke discharged for about 1.5 million minutes each day![74] In industrial areas, the death rate from respiratory diseases was about eight per 1,000. In some of the residential areas, with no mills, it was 17 per 1,000, jumping to about 30 per 1,000 when the rains failed or the wind died down.[75] Domestic smoke, it was clear, had much to contribute to air pollution and to respiratory diseases, the single-largest cause of death in Calcutta:

> In Calcutta and the suburbs, during the cold season, when the atmospheric conditions trap the smoke near the ground level, when curtains and clothes are blackened in a few hours, and the nostrils and the respiratory tract are clogged with soot, the death rate from diseases of the breathing organ rise from 80 to 240 a week . . . The smoke nuisance is ancient . . . [and for long] there seemed to be

> no commercial solution ... The scientific use of fuel solved economically, at least, the problem of industrial smoke abatement. But domestic smoke, which is responsible for about 75 percent of the damage, is not yet commercially solved. Under present economic conditions, the effect of cheap and smokeless power and heat is profound and is closely related to the health and prosperity of a nation.[76]

With this, the Bengal commission signalled a definitive move away from consideration of smoke exclusively as an engineering and efficiency issue (see Chapters 3 and 4) to one that needed to focus far more on health.[77] It was imperative now that the department be reorganised, bringing on board health specialists who could assist the commission.[78] Dr C. A. Bentley, director of public health, Bengal, was thus included in the commission beginning 1927, and in the following years, the commission spent considerable energy on elaborating its smoke/health agenda. Investigation were started following prevailing medical opinion that in large congested towns, the death rate from breathing organ diseases was a fair working index of the degree of atmospheric pollution by smoke.[79] Wind, the commission explained, was the natural scavenger of the atmosphere, with a velocity of 11–12 miles an hour sufficient to keep a town moderately clean of smoke. However, when it dropped to 2–3 miles an hour, as it did in Calcutta the winters, the scavenging failed, leading to the air thickening with impurities and an increasing death rate on account of respiratory problems.[80] Rain was the other natural scavenger, but in Calcutta, and elsewhere in India, it was confined to a few months only.[81] The air being breathed in the city, in other words, was both polluted and stagnant and, consequently, unfit for human consumption, at least in some districts. Especially in the winters, Calcutta was 'gassed'.[82] And its numerous harms found elaborate mention in the annual reports of the commission—smoke devitalised, disabled and poisoned; it made the linings of the lungs rough and thickened, causing chronic bronchitis. The irritation could also cause a partial collapse of the lungs, the incomplete oxygenation of the blood even leading to tuberculosis. Acute respiratory diseases caused

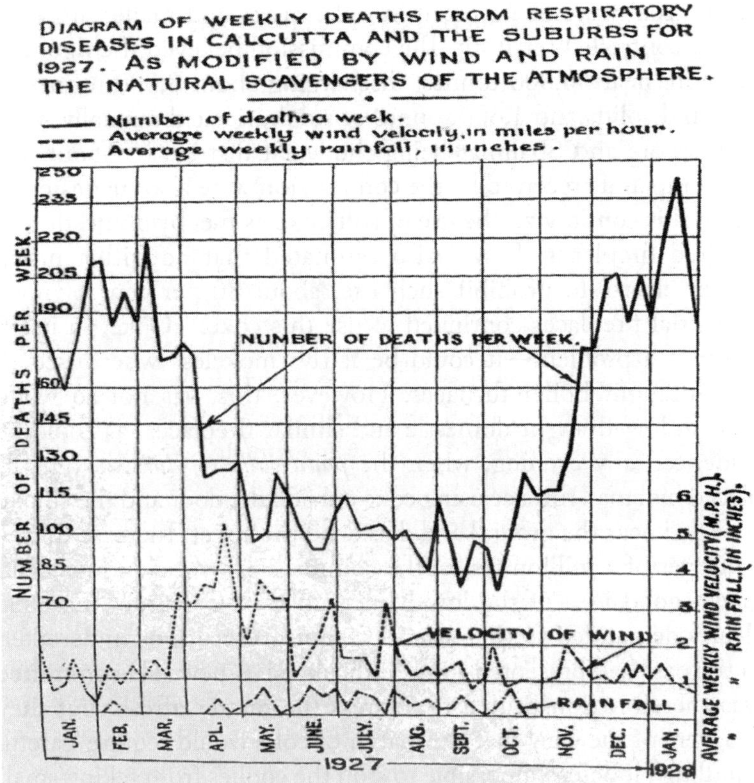

Source: Bengal Smoke Nuisances Commission, *Annual Report*, 1927, Appendix 3. The explanation of the diagram, provided in the report, is as follows: 'It appears that in the first week of Jan 1927, with a wind velocity of nearly 2 miles per hour that the death rate is 186. The wind then falls and the death rate rises. About the middle of March wind velocity increases and the death rate falls. About June, the rains of the monsoon start, which wash the pollution from the atmosphere. The death rate gradually drops to 77 a week in October, when the rains stop. The wind velocity drops to about 1 mile an hour and the death rate rapidly rises, ending with 244 deaths a week in January 1928.'

Figure 5.3: Wind: The Natural Scavenger of Air

more deaths than all other infectious diseases combined. City fog was 'a mass murderer'.[83]

Attention again turned to fuel use, especially the use of coke. At its most extreme, some estimated that about 1,00,000 tonnes of raw coal were being burnt in open heaps and made into coke in the controlled territory. The commission believed the figure to be more near 50,000 tonnes, discharging about 3,000 tonnes of soot and solids and 1,600 tonnes of sulphuric acid annually, with Kidderpore and Shalimar being the worst districts.[84] About half the complaints received by the commission were also suggested to be on this count, viz., the use of soft coke as fuel by domestic and kindred fireplaces. It was also estimated that notwithstanding earlier efforts to prohibit such use, about 80 per cent of non-industrial fireplaces continued to use this coke.[85] Coke, in itself, was not a problem—it could be fairly smokeless when used in high-draught boiler furnaces. However, this was not so when used in low-draught domestic and similar fireplaces, as could be evidenced any evening, when the *paanwalla* or *durwan* (guard) placed his small furnace using coke outside the door and the smoke travelled near the ground and darkened the street. Increase this by a quarter of a million times, the commission wrote, and it would give a good idea of the low-lying smoke in Calcutta. The issue clearly demanded further attention, but prosecutions undertaken in the past had presented some 'difficulties'. A new sub-committee was therefore constituted to address the matter afresh, but they too were of the view that the matter of coke would require 'careful handling', it being impossible to stop the coolies from taking small quantities of coal for their domestic use. This may well be, the commission reckoned, but the manufacture of soft coke for sale must be entirely stopped, commending the action taken by the Calcutta port commissioners in this regard.[86]

Some success was to be had it seemed, the commission reporting in 1930 that 'for some time, the air of Calcutta has been cleaner. There have not been so many smoky nights. The smoke seemed less concentrated and not so irritating to the respiratory tract. Under such conditions, respiratory deaths should decrease.'[87] Some success was to be had regarding brick kilns too, especially in the Tollygunge area, which had given rise to many complaints from the residents of the locality. In all, there were over a 100 brick-fired chimneys in the controlled territory, working for a few months, discharging about 8,000 minutes of black smoke during such months. So far this had been an infinitely small figure compared to about 30,000 minutes of smoke being discharged daily by factories and therefore of no great practical significance. But in view of the control that had been achieved with respect to factory smoke, something could now be done about brick kilns too.[88] The problem though was that the bricks were burnt in sections in circular trenches with short chimneys put up. When the bricks in one trench were ready, the chimney was removed to the next section and so on, until all the contents of the kiln were completed. Consequently, brick kilns could not have fixed high chimneys, leaving the removal of kilns from the locality as the only option for dealing with their smoke. The government was happy to endorse this proposal, issuing a notification in November 1932 prohibiting the erection or use of brick kilns within the local limits of the South Suburban Municipality, unless such kilns were provided with chimneys approved by the commission and were not less than 100 feet high.

The matter of domestic smoke, however, remained to be tackled effectively, with the director of public health noting the continued use of cow dung and soft coal instead of coke, not only in the homes of the poor but also in the mansions of the upper and middle classes.[89] The consequence, of this and smoke from other sources, once again was the lodging of fine carbon particles inside the lungs, leading to irritation of the throat and, in some instances, even more acute diseases like broncho-pneumonia and pulmonary tuberculosis. In the specific instance of smoke from domestic fires, there was also the potential of irritation of the conjunctiva, which under chronic conditions could lead to severe eye ailments, including blindness.[90] In response, the commission was yet again reorganised, with the chief and assistant inspectors visiting the by-lanes every morning and evening and carrying out instruction tests, demonstrating to the people how more draught when lighting their fires could be a key to abating smoke.[91] Early results were again believed to be positive, with only eight months of work under the reorganised system enough to convince the commission that domestic smoke could be greatly reduced to the satisfaction of the general public. Given this, and despite inadequate resources, it was imperative therefore that the commission continue to extend its activities, even as it worked towards possible amendments in the Act to include domestic smoke.[92] Indeed, the commission was willing to go so far as to suggest that no plan should be sanctioned for dwelling houses unless a serviceable chula was installed and that water and sewer connections be refused till full compliance with the sanctioned plan had been demonstrated![93] Such an amendment was not be had, however, and the commission had no option but to regulate domestic and kindred smoke as best as they could through popular demonstrations of good firing practices. Indeed, not much progress could be made as regards fuel either, and well into the 1940s, it continued to be observed that large quantity of coke and coal continued to be burnt in towns and villages for kitchen fires. So long as domestic smoke continued to be beyond the Act, any possibility of the reduction of this waste and 'the useless burning of fuel in kitchen fire' remained distant, even though it was emphasised that such a reduction would be of great national, industrial and economical importance.[94] Smoke, the commission was forced to acknowledge, remained very much 'dirt in the wrong place', with domestic smoke 'the most dangerous kind of smoke'.[95]

Bombay was to fare even worse in this regard. Indeed, it would seem that right from the year of its inception, the Bombay Smoke Nuisances Commission had little to say on the issue of domestic smoke. As in Calcutta, there was definitely a recognition that domestic smoke was no less responsible for vitiating the atmosphere than industrial smoke, but this was typically followed by the qualifier that the scope of the Act was only confined to industrial smoke, referring the matter of domestic smoke to the health officer.[96] In the instance of a complaint from Mr D. P. Bett regarding the nuisance caused by the emission of smoke from one of the domestic chimneys of Thoburn House, Apollo Bunder, thus, the complainant was informed that the Smoke Nuisances Department had no jurisdiction over domestic chimneys, and accordingly advised the complainant to address the Bombay municipality in this matter.[97] A similar response was given to Messrs Kiddle Reeve and Company, 21, Elphinstone Circle, Fort, regarding a complaint against a domestic chimney in a private residence at the back of their balcony. The next year (1918), two minor complaints relating to domestic kitchen chimneys were again received and met with same response—as these were beyond the scope of the Act, they were forwarded to the health officer of the Bombay municipality for disposal.[98] Occasionally, the issue could concern both, the commission and the municipality. Thus when Mrs A. Ring Archer complained about soot and smoke from the chimney of the Apollo Hotel, 'the matter was set right so far as it related to the Smoke Nuisances Department and the complainant was referred to the Bombay municipality for the redress of other grievances'.[99] Health reports of the municipal commission, however, had little to offer on the subject.[100] Some attempt was made to reach out to members of the public, the chief inspector attending one of the meetings of the local Sanitary Association where he impressed upon them the need to focus on the abatement of domestic smoke since this was not covered under the Act and yet was responsible for smoky evenings in many months of the year.[101] But in the main, the only viable option now seemed to be the greater use of gas and electricity, together with wider publicity. The Smoke Nuisances Commission would not be alone in singing praises of what electricity could do inside homes either, being joined in by members of the public too:

> The average middle class Indian home is small and generally without proper drainage and ventilation, in which kitchen smoke is a characteristic feature ... The introduction of cheap electric power will change all this and its most beneficent and humanising effect will be on the Indian homes, which it will make cleaner and more comfortable ... when electricity displaces firewood there will be no smoke nuisance.[102]

Manufacturers and commercial establishments would not be far behind either, advertising how the advances made in the industrial sector could be transferred within homes too. 'The energy that launches great liners can be used for your Household

Source: *The Bombay Chronicle*, 10 Feb. and 6 Mar. 1937.

Figure 5.4: Electricity: The Eternal Hope

Refrigerator', read the tag line in one advertisement.[103] 'The power that sends trains across mighty continents can heat your water', read another.[104]

And so it was to the use of electricity that the question of domestic smoke would turn.

GAS, ELECTRICITY AND PUBLICITY

To reiterate, the promise of electricity as the ultimate solution for the abatement of industrial smoke had been intimated since the early years of the twentieth century. Similarly, given the many complaints against domestic smoke, about which the commissions were helpless in doing much, emphasis was laid on the use of gas and electricity within homes.[105] Aiding the process was the attempts made by the gas company in Calcutta to open itself to public scrutiny of its heating and cooking appliances and to instruct them in the use of such appliances. As a result, 100 coal-fired furnaces were reported to have been replaced with gas heaters in 1909–10, while 50 gas cookers replaced the ordinary smoky domestic coal-fired cooking apparatus.[106] In addition to this, some of the leading restaurants were also reported to have fitted gas cooking apparatus, replacing those that utilised coal. As yet, such substitutions in favour of electricity, oil and gas still seemed driven more by convenience, the commission noted, though the avoidance of smoke trouble had 'also an influence in the matter'. The efforts of the gas company to reduce prices and enlarge their capacities were also seen as hopeful signs, especially in the extension of its use beyond cooking and heating to include lighting, thus doing away with the objectionable oil lamps.[107] Prices were reduced in 1914–15, and the commission reported that a larger use of gas-worked appliances for water heating, cooking and lighting were installed even during the war years when there was a scarcity of materials.[108] The post-war years saw even more advances being made in this direction, with the following gas appliances installed in Calcutta by the Oriental Gas Company (OGC) in 1919–20: private cooking ranges (262), small cooking stoves (8), large cooking ranges for hotels (2), hot closets for plates, etc., (2) and geysers for heating water (130).[109] The next year, attention was again drawn to the installation of gas cookers in large numbers, and it was expected that when their advantages were more widely known, their adoption would be further accelerated.[110]

Source: *The Statesman*, 14 Jan. 1917. 'With these wonderful stoves', the advertisement reads, 'cooking can be done in your own room, verandah, or wherever you wish, for there is neither smell, smoke, nor dick' [*sic*.].

Figure 5.5: Gas: The Antidote to Smoke

Much was still by way of experimentation, even regarding industrial furnaces, but there was little doubt that there was a wide field of enterprise emerging in the application of smokeless gas appliances with newer items finding a place within homes and commercial establishments—cookers for domestic use and for hotels, gas-heated circulating boilers for heating water, similar boilers for tea-making in hotels and restaurants, and hot cases and coffee boiling machines for restaurant use.[111] And as raw coal steadily replaced coke in domestic fireplaces, the installation of gas appliances assumed even greater importance.[112] Large residential buildings being built were seen as especially lucrative sites for the same, the OGC making special mention of two large buildings on Park Street in which 82 gas cooking ranges and 76 gas geysers were installed in 1922.[113] Slightly over 1,000 gas appliances for domestic purposes were added the next year and an equal number the year after, with further efforts being made to promote the use of gas appliances in hotels, boarding houses, clubs and new pucca buildings.[114] The scale of transformation was reportedly impressive—there had been no more than 100 gas appliances in use when the commission had started its work, but within a decade and a half, the number had reached nearly 10,000, preventing 'the annual discharge of about 1,800 tons of soot and 700 tons of sulphuric gases into the air of Calcutta'.[115]

Further advances, however, would depend on pricing, fuel substitution within homes being as price-sensitive as in industry. Cheap gas was an absolute necessity for Calcutta, the commission wrote, and while 10,000 appliances was a good number, the need was for a quarter of a million.[116] And for this, it was absolutely essential that the gas company reduce its price, something that was eminently possible if the municipal corporation could be persuaded to switch from lighting gas to heating gas.[117] The corporation was not easily persuaded though, their contract with the OGC being for the supply of gas of illumination or candlepower standard. This, the commission believed, was hugely expensive, pointing out that with the introduction of glass mangles, towns across the world had begun to move towards the use of heating gas, which was both cheaper and more efficient.[118] Gas engineers the world over were also increasingly collecting the gas from the making of coke in industrial gas ovens, sometimes from considerable distances, and mixing it with town gas to reduce the price, in a bid to extend use.[119] Indeed, in India itself, Bombay had shifted to heating gas much earlier. The objection could not be to the gas in question either, for the new heating gas had been in use throughout Calcutta for two months in 1928 without anyone noticing the difference.[120] And so, the commission opined, 'the obsolete restrictions that the corporation compel the gas company to observe' must go. A meeting was organised between members of the commission, the corporation and of the gas company where the gas company offered to reduce Calcutta's annual gas bill by Rs 1,26,000—the corporation benefitting by Rs 28,000 and the public by Rs 98,000—with effect from 1 January 1928, provided the Calcutta corporation would agree to change over to the calorific standard. This was unanimously accepted. The case was then passed on to the Calcutta corporation for consideration where the matter seemingly rested for a couple of years. Eventually in December 1931, the corporation agreed to make the switch, but by then a loss of over Rs 5 lakh had been incurred.[121] Others too chimed in, the director of public health writing that if public utility companies that produced gas and electricity could be induced to make these available to the public at a comparatively cheaper rate, then the extended use of these fuels would yield 'a very valuable result in diminishing the volume of smoke'.[122] This was not all. As we saw before, around the same time, experiments were being made to produce smokeless fuel such as gas and oil from raw coal through low-pressure carbonisation. The electric and gas corporations, being situated in proximate distance, were quite suited for this purpose. The electric supply corporation could process the raw coal, burn the semi-coke in the boilers, dispose of the oils and deliver cheap gas in bulk to the gas company, which could then distribute it to Calcutta, producing a win-win situation—large reduction in fuel charges, and consequently, cheaper electricity and gas for the city, greater incomes for electric and gas corporations, and perhaps most importantly, ensuring decline in mortality.[123] The dreams of a smokeless city were much alive: 'When a smokeless fuel is produced that can be thrown anyhow into a furnace without producing smoke, it will remove the "human element", the last barrier to the realisation of the smokeless city.'[124] However, further consideration of this in Calcutta would have to remain suspended, pending a decision regarding the heating standard for gas that would be adopted in the city.[125] Meanwhile, the company itself reduced the heating rate from 1.25 annas per unit to 1 anna per unit, leading the commission to hope that there was nothing to prevent a more extended use of gas and electrical energy for domestic purposes such as cooking, heating, refrigeration and all other activities incidental thereto.[126]

By the end of the 1920s, results on both these fronts seemed encouraging. The number of gas appliances, including geysers, cookers, etc., steadily increased by over 1,200–1,300 each year, and the commission was content to observe that compared with the cheapest liquid or solid fuels, its advantages from a health point of view alone made it preferable for cooking and heating.[127] The efforts to popularise the use of electricity for cooking and other domestic purposes were also yielding results, the commission noted, especially in the use of hot plates, grilles and toasters.[128] Indeed, the Calcutta Electric Supply Corporation (CESC) was looking to develop this further, being convinced of the advantages of electrical cooking, obviating the dirt, smoke and fumes inevitably connected with any other form of heating.[129] An interesting history of how the company had reached here was provided by the enquiry into electrical rates conducted by the government in 1936.[130] The company, which began in 1897, saw an immediate run on electric fans, as well as a steady growth in the demand for house lighting. The demand for electric energy to be used for ventilating and power purposes, as well as for lighting, was, at one level, anticipated, but no one had quite imagined that the days of the hand-pulled punkah were numbered and that the electrically driven fan would soon take over. But such indeed had been the case and the popularity of the electric fan ensured the immediate prosperity of the company, providing, at least during a great part of the year, the 'day load' that was so essential to the economical working of an electricity supply station.[131] Some apprehensions were expressed about whether the transfer of capital to Delhi would reduce demand, but if anything, the enquiry noted, there was a steady increase in demand for electricity for residential purposes, people even thinking it fit to light temples with it![132] Rates for domestic consumers had also tended to come down over the years, from 3.5 annas per unit in 1925 to 2.5 annas per unit in 1935. However, in view of the fact that the load for domestic consumers was the foundation on which the company's prosperous undertaking had been built, the committee reported, it was only fair that they share in the prosperity of the company, suggesting therefore to lower the rates further to 2 annas per unit, which, they believed, would spur the demand even further.[133] Some limits still remained though. There was little or no demand for general heating as water heating could be done at very low cost by coal, coke or other wood fuel and washing services could be obtained at a reasonably monthly cost with no initial capital expenditure. Refrigeration was also beyond the reach of the large majority of consumers. Similarly, not only was the initial cost of consumers' heating and cooking appliances high but also a satisfactory cooking appliance suitable for Indian requirements that could be obtained at a cost within the purchasing power of the average consumer had not yet been placed on the market.[134] But the enquiry remained optimistic. This too will soon happen, they wrote, while recommending that rates be brought down from 1 anna to 0.75 anna. The commission too shared in this optimism, as indeed it had from its very inception, writing yet again that once electricity was introduced in any setting, it could be used for light, heat and power in a diversity of ways that no other form of energy could equal. And once this were to come about, the general use of electricity would 'go far to dispel those clouds of smoke that hang over Calcutta during the cold weather evenings'. Similarly regarding gas—'compared with the cheapest liquid or solid fuels its advantages from a health point of view alone make it preferable and for cooking and heating its supremacy cannot be successfully challenged'.[135] However, while it was one thing for the suppliers to be convinced, it would be another matter altogether to persuade consumers to invest in these gadgets.

If price was important, so was publicity. The peculiarities of domestic smoke meant that a much closer relationship had to be forged between the commission and the public. As we saw earlier, one of the measures to achieve this was the reorganisation of the commission to enable members to travel to the localities in the company of the chief inspector and to interact with the public there. Beginning the mid-1920s thus, publicity became an important plank of the commission's activities. A system of issuing bulletins on points of public interest was started in 1927, and the next year, the Bengal commission showcased practical exhibits, showing the effect of smoke nuisance and air pollution on mortality, plants and food supplies, buildings, distant localities and sunlight.[136] It also began the practice of controlling a section in the Calcutta Health Welfare Week Exhibition to exhibit the same.[137] Alongside, it also collaborated with the OGC, the latter showcasing smokeless gas appliances for cooking and other domestic purposes.[138] Further measures were adopted in 1932, including printed copies of illustrated instructions being forwarded to the health officer of the municipal corporation for distribution among the public through ward health associations and arrangements being made with the cinema authorities to screen films on the evils of smoke and how to abate it.[139] The results seem to have been a tad disappointing though, the commission observing in 1933 that despite their collaboration over three years with the Health Department and various ward health associations 'to use persuasive and educational methods to minimize this source of evil', little headway had been made.[140] In this, they may not have been too off the mark, the municipal corporation's own report on publicity being carried out by the newly created department for spreading knowledge about health and hygiene reporting on lectures and use of posters and handbills as regards small pox, cholera, tuberculosis, typhoid, malaria, general sanitation, personal hygiene, maternity, etc., but to the exclusion of smoke nuisance.[141] In 1938, 39,356 lectures were delivered, 23 varieties of handbills and posters printed and circulated, health stalls organised by the ward health associations in different exhibitions and the American method of arranging stall adopted for the Annual Health Week Exhibition. Yet not one mention was made of any specific anti-smoke propaganda.[142] Infectious diseases and personal hygiene, it would seem, continued to remain the focus as it had been from the very inception of the municipal health department.

Notwithstanding, the commission continued to have faith, being convinced that the various health exhibitions held in the city was awakening at least some interest in the public in the issue of clean air. Evidence that property owners and occupiers were increasingly realising the benefits and advantages of the use of gas as a smoke-eliminating fuel could be had from the fact that despite the continued prevalence of depressed business conditions, nearly 600 new gas-using appliances had been added in 1933, comprising cookers, grilles, hot plates, ring burners, toasters and other domestic appliances, while appliances for industrial purposes using gas included steam-raising boilers, metal melting and hardening, furnaces, drying ovens, Tinman's stoves and soldering machines. The next year too, the domestic demand for gas came in for special mention, with new residential buildings being fitted with complete installations for cooking and water-heating purposes, the demand for gas geysers for the latter function being especially marked.[143] And to demonstrate that their endeavours were not addressed to the middle and upper segments alone, posters were distributed among ordinary persons at the Calcutta Health Week demonstrating ways of abating smoke with extra draught. The objective was clear—those who could afford should install electric or gas heating appliances and chimneys, and those who could not should adopt the simple method of fanning the fire strongly for three or four minutes when being lighted, thus eliminating smoke.[144] Press and pamphlets were not alone either, with health and smoke abatement authorities reaching out through other media platforms such as the radio, with a broad-cast by the deputy mayor of Calcutta in 1941 on the persistence of domestic smoke in the wake of increasing population and increasing number of bustees.[145] Once again, the appeal was to all public institutions interested in the health of the city to assist them in propaganda and demonstrations, and thereby help reduce the number of deaths from respiratory diseases, which were greater than the total from any other two single diseases during the cold season.[146]

The Health Week and commission-related publicity were not the only tasks undertaken by the gas and electricity companies. Looking to attract new consumers they put out advertisements of their own, with electric companies focusing on fans, lights, etc., while the gas companies targeted the space of the kitchen. Advertisements in the press, Sarkar writes, was complimented by way of handouts made of blotting paper, with messages inscribed on them, and door-to-door campaigning, such canvassing continuing well into the 1940s.[147]

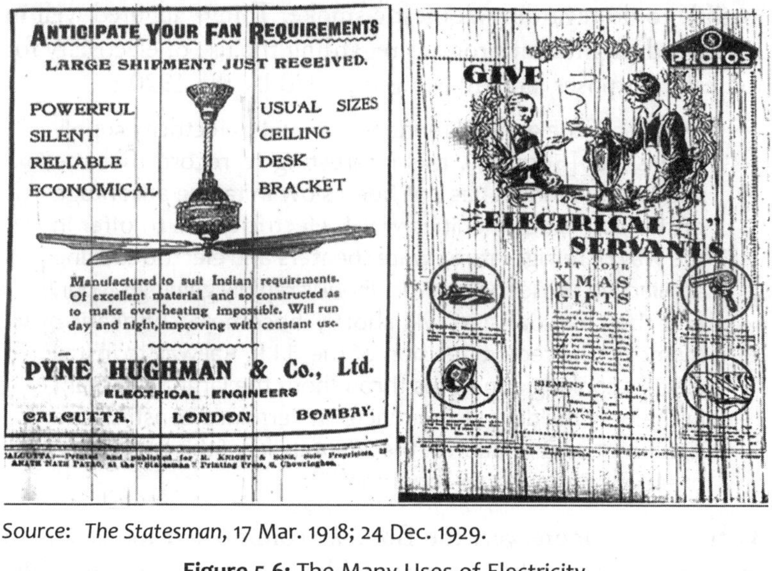

Source: *The Statesman*, 17 Mar. 1918; 24 Dec. 1929.

Figure 5.6: The Many Uses of Electricity

However, in this success would also lie some of the limits, the OGC reporting in 1944 that they could not connect any additional appliances to their mains during the year as the consumption of gas for industrial and domestic purposes had increased to such an extent that their manufacturing plant and distribution system was loaded to full capacity.[148] As for the CESC, while the area served by them had close to 1.5 million people, merely 57,415 houses were connected with electrical wiring by 1940.[149]

In Bombay too there was an argument developing by 1915–16 that the health of the city had nearly reached 'Western' conditions, affording authorities the opportunity to turn their attention away from infectious disease such as plague, small pox, cholera, malaria, etc., towards dealing with diseases that, although not so infectious and fatal, were the cause of much sickness and suffering.[150] Domestic smoke, however, failed to find much address in the activities of the commission as yet, which, as we saw earlier, concentrated its energy on the small industrial establishments. So even more than in Calcutta, the use of electricity and gas, and the publicity around it by private companies, would form the core of what was possible to do regarding domestic smoke. Things augured well in this regard, significant progress regarding the use of electricity for domestic purposes already being reported by the 1920s:

> As regards the general tendency of public electricity supply in this city [Bombay], it is interesting to record that the middle-class Indian has not been slow in taking advantage of the domestic amenities which electricity has to offer in the way of refrigerators, water heaters and electric cooking appliances, while the electric *dhobi* is rapidly gaining favour, and the charcoal *sigri* will shortly become a suffocating memory. The recent decision of the G.I.P. Railway to install electric cooking apparatus throughout the large restaurant in the new building at their Victoria Terminus constitutes a significant confirmation of the adaptability of this application of electricity to Indian conditions.[151]

By the 1930s, moreover, the Bombay commission too seems to have realised the value of such publicity, informing the public about the availability of electricity in the Town and Island of Bombay, encouraging its greater use, while electric companies advertised this being available at reduced rates.[152]

Hotels and restaurants were advised to make greater use of gas and electricity and thus avoid potential complaints from neighbours.[153] The actual use of electricity and gas was also noted, the

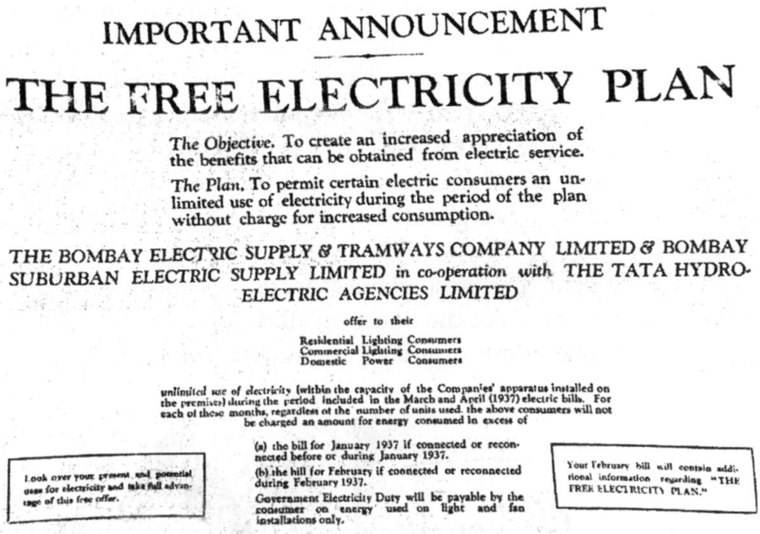

Source: *The Times of India*, 25 Feb. 1937.

Figure 5.7: The Free Electricity Plan

commission observing the marked improvement that these had made in residential areas.[154] 'It has been noticed', they wrote in 1933–34, 'that domestic cooking is now being done on an increasing scale by gas and if this method is adopted much of the domestic smoke will be done away with to the great improvement of the general health of the city.'[155] However, constraints remained. As in Calcutta, a propaganda officer had been appointed by the Bombay municipality too, to educate the public in matters of public health with the help of latest propaganda material such as slides, films, posters, lectures, etc., but even here, the public outreach did not include matters of smoke and health.[156] The commission repeatedly urged the public to use coke or smokeless fuels such as gas and electricity but did not combine with either the municipality (as in the Calcutta Health Week) or with industry, leaving the latter to advertise their wares rather independently of the commission's activities. The advertisements were varied and interesting too in the narrative they sought to build. Electric companies, not surprisingly, focused on the good that each particular item could provide, lights and fans continuing to be clear favourites. But they also went beyond this, drawing the consumers' attention to an entire range of functions that could be delivered far better through the use of electricity. As one advertisement put it: 'Are you using electricity? Or Only electric light?' Often the emphasis would be on health and comfort, though occasionally the specific fact of enabling cooking without flame or smoke would also find mention.

Advertisements for gas appliances were even more pedagogic, instructing the woman of the house on all the potential benefits, in terms of costs and comfort, indeed in terms of becoming properly

Source: *The Times of India*, 20 Jan. 1937; *The Bombay Chronicle*, 7 Apr. 1934.

Figure 5.8: Electricity: Its Multiple Benefits

urban herself! Occasionally, they would also speak to getting rid of smoke too. As the lady in the picture in the centre declares, she always prefers to cook with gas as 'It is cheap ... It is clean ... no smoke, no soot ... just a clear, bright flame' (Figure 5.9).

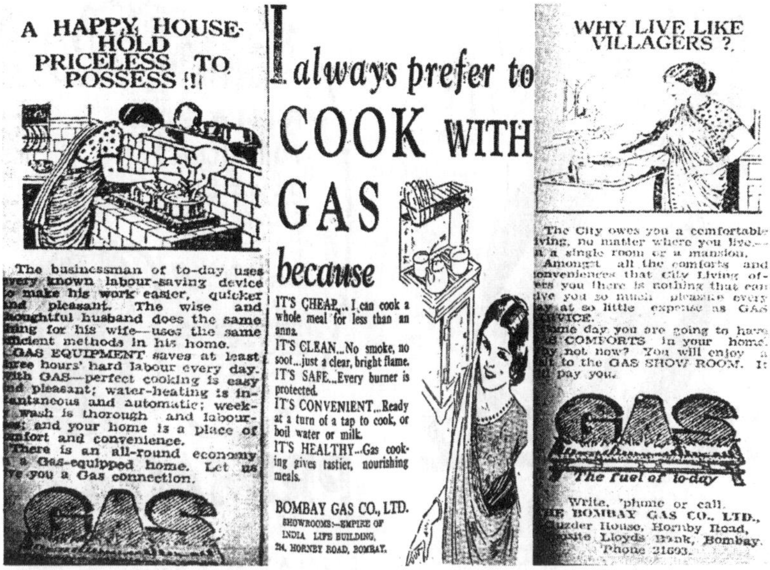

Source: *The Bombay Chronicle*, 25 and 27 Dec. 1930; *The Times of India*, 17 Sep. 1934.[157]

Figure 5.9: Gas: Cheap, Convenient, Healthy

And as bottled gas became available, making possible its use both indoors and outdoors, the figure of the woman interestingly changed with a more Westernised woman and her family emerging as the proper subjects of the enjoyment of a picnic! Once again, among the many benefits would be 'no smoke or dirt and nothing to go wrong'.[158]

Quite evidently, the question of just who and how many among the residents of the city could make such a transition went unasked.[159] Any reduction in smoke, through the adoption of better fuels and new technologies, it would seem, was a task to be performed by only a miniscule minority of the middle class. Not surprisingly, the burden of smoke within households,

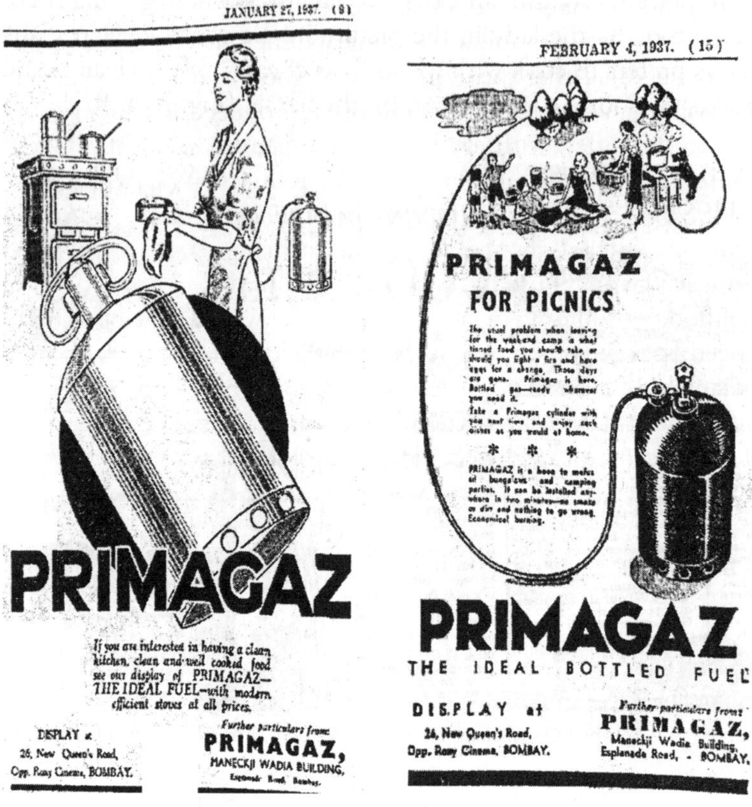

Source: *The Times of India*, 27 Jan. 1937; 4 Feb. 1937.

Figure 5.10: Picnic with the Comfort of Gas

painstakingly documented by labour activists and health officials from the first decades of the twentieth century, could find little meeting ground with domestic smoke as an urban nuisance, which had come to be a major preoccupation of commissions from around the same time.

Conclusion

The question of domestic smoke, we have suggested in this chapter, was imbricated in two different concerns—on the one hand was the question of labour and the amenities offered to Indian workers, especially by way of sanitary housing. Numerous enquiries on this count suggested that the availability of good ventilation was near impossible, and domestic kitchens often produced smoke that had little to no outlet, with every available opening being shut off to ensure greater availability of space. Conditions in Calcutta were possibly better on this count than in Bombay, though even there to the extent that cooking was done in living spaces with wood and other biomass used as fuel, in addition to (inferior) coal, the likelihood of suffering on account of smoke would have been high. Indeed, as some reports suggest, given the design of the traditional Indian chula, even middle-class homes would have suffered on this count, though some alternative designs also emerged. And through all this, accounts vacillated between holding habits and cultural values of the working poor responsible for the harms that occurred on account of smoke and other nuisances, or attributing these to labour–capital relations that failed to provide for decent wages and hence also the failure to provide decent housing.

On the other hand, there was a more generalised concern of the Smoke Nuisances Commissions that worried that domestic smoke, being emitted even closer to the ground in the absence of chimneys in Indian homes, was as much of an urban issue (affecting the city at large rather than individual homes alone) as industrial smoke. Indeed, by the middle of the 1920s, when industrial smoke seems to have been controlled somewhat, the commissions increasingly turned their attention to the health impacts of domestic smoke, viewing it as a bigger public health concern than industrial smoke. Calcutta, however, seems to have worried far more on this count than Bombay, offering detailed analysis of the manifold links between domestic smoke and the harms to health and property. Bombay, by contrast, seems to have worried far more about smoke from small-scale industrial establishments, which too, being often located in dense residential neighbourhoods, were a source of nuisance. However, neither in Calcutta nor in Bombay was the Act amended to include domestic smoke in its ambit, leaving the utilisation of gas and electrical appliances within homes and in commercial establishments as the only available remedy. Manufacturers were keen to market these for commercial reasons, while the commissions were keen to propagate them for their own ends. For the companies, absence of smoke would feature as one among the many advantages that they offered. For the commissions, it was only this dimension that was of essence, highlighting these in their many publicity initiatives. Ultimately, however, in both Calcutta and in Bombay, there was a clear recognition that the extent to which these could be promoted was highly differentiated, being possible only in larger middle-class homes and among the more recently constructed flats. Smoke would thus simultaneously be an urban issue and the burden of individual households, with effective responses to it configured in individual terms. Not quite 'green consumerism' but certainly indicative of the different scales on which the question of smoke was articulated in colonial India, to be effectively countered through the use of new energy sources.

NOTES

1. Letter by a Belgian priest, 1867. Cited in Nilanjana Gupta, ed., 'Introduction', in *Strangely Beloved: Writings on Calcutta*, Delhi: Rupa Publications, 2014, p. 7.
2. Bengal SNC, *Annual Report*, 1927.
3. There were eight general industrial strikes in Bombay between 1919 and 1940. In Calcutta, major waves of strikes were witnessed in 1920–21, 1929 and 1937–38, many of these inspired by communist trade unions. The figure is even more staggering when looked at the level of individual firms, with 89 strikes in the Calcutta industrial area in the last six months of 1920 alone. See Bengal, *Report of the Committee on Industrial Unrest in Bengal*, Calcutta: Bengal Secretariat Press, 1921, p. 1.
4. MSA, File No. 1311, 1920, Pt. 1, Box 314. To this, the sanitary commissioner of Bombay responded that improvement would lead to higher values, which, in turn, could lead to the displacement of the poor. Hence any act of improvement must be accompanied by town extension programmes. 'Development must precede improvement', he wrote, 'and march always slightly in advance of it.'

5. As an aspect of this, far greater attention was paid than before to the lives of workers, their sexuality, crime and diseases. See Samita Sen, *Women and Labour in Late Colonial India: The Bengal Jute Industry*, Cambridge: Cambridge University Press, 1999.
6. We may recollect the sentiments of industrialists such as N. N. Wadia who had argued for properly constructed, cheap and well-ventilated buildings, under proper supervision as to cleanliness and sanitary requirements. Similar sentiments were expressed in the press too, the *Indian Spectator* writing that if there was one thing that the plague had most impressed on the public mind, it was the necessity of having the labourers 'employed in the great industrial enterprise of the city', 'housed well and in healthy surroundings'. Cited in Priyanka Srivastava, 'Creating a Healthy and "Decent" Industrial Labour: Health, Sanitation and Welfare in Colonial Bombay, 1896–1945', Ph.D. thesis, University of Cincinnati, 2012, pp. 76–77.
7. India, *Report of the IFLC*, Vol. II, Simla: Central Branch Press, 1908, p. 67. There were other mill owners though who continued to speak in favour of them building such chawls. See, for instance, the testimony of the manager of the Dinshaw Petit Mill in the same report, pp. 81–83.
8. *Report of the IFLC*, Vol. II, pp. 238, 245. Mill owners also differed in their assessment of whether the workers themselves liked to live in company-owned chawls or preferred private housing.
9. And by workers they implied not only mill workers but also dock and godown labourers, employees of the railways and other public bodies, and coolie labour. See GoI, *Report of the Indian Industrial Commission 1916–18*, Calcutta: Superintendent Government Printing, 1918, pp. 180–89.
10. Henrik Aspengren, 'Empire: A Question of Hearts? The Social Turn in Colonial Government, Bombay, c. 1905–1925', in Mark Duffield and Vernon Hewitt, eds, *Empire, Development and Colonialism: The Past in the Present*, Woodbridge, Suffolk: James Curry, 2009, pp. 43–58. But as the *Bombay Chronicle* was to remind, it was one thing to show compliance with international obligations, quite another to actually improve the living condition of workers. See 'Factory Labour in Bombay Presidency', *BC*, 11 Nov. 1930.
11. In Bombay, this took the form of a competition in 1918 to design a ground floor or two-storey building, no tenement in which should have less than two rooms and the total cost of which would

not exceed Rs 750, the best chosen design then being made available to the public for constructing their own homes. See A. E. Mirams, *Plans and Specifications of Houses Suitable for Occupation by the Working Classes*, 2nd edition, Bombay: Government Central Press, 1925. The first edition of these plans was published in 1919 and was reported to have found an unprecedented large demand, with requests for publications being received from all parts of India.

12. Nikhil Rao suggests that the BDD was created to oversee the orderly development of the city in a time of economic growth, which proved to be rather short-lived. Towards this, it was assigned three distinct tasks: (i) reclamation of land in the foreshores to the west, (ii) construction of single-room tenements in the industrial districts (the Industrial Housing Scheme) and (iii) the development of suburbs on Salsette Island. See Nikhil Rao, *House, But No Garden: Apartment Living in Bombay's Suburbs, 1898–1964*, Minneapolis: University of Minnesota Press, 2013, p. 110.
13. The revised figure was 16,000, and of these, 12,000 went unoccupied.
14. Government of Bombay, *Report on the Working of the Development Directorate for the Year Ending 31st March 1924*, Bombay: Government Central Press, 1924. Rao instead suggests that the real failure of the BDD lay in not recognising that the real demand for housing came not from the working class but from the upper-caste, lower-class, white-collar workers. See Rao, *House, But No Garden*, p. 111.
15. For a detailed account of housing density in Bombay, see Rajnarayan Chandavarkar, *The Origins of Industrial Capitalism in India: Business Strategies and the Working Classes in Bombay, 1900–1940*, Cambridge: Cambridge University Press, 1994, pp. 175–88.
16. 'Houses for Workers', *BC*, 13 Jun. 1934.
17. Partho Datta, *Planning the City: Urbanization and Reform in Calcutta, c. 1800–c.1940*, Delhi: Tulika Books, 2012, p. 269.
18. Cited in Datta, *Planning the City*, p. 271.
19. This is what had prompted them to make a design innovation, providing for reinforced cement concrete (RCC) lattice instead of a regular window, which would allow for air to come in while keeping the sun out. However, class and caste prejudices may also have been at play here, with some BDD officials suggesting that while windows could well be provided for Muslims, Christians, Jews and high-caste Hindus, this other arrangement must continue for lower-caste and lower-class residents until they had been 'educated

to use a window as a means of fresh air and not as an ever present outlet for refuse, both solid and liquid'. See Rao, *House, But No Garden*, p. 114.

20. Rao, *House, But No Garden*, pp. 114–15.
21. Dipesh Chakrabarty, *Rethinking Working Class History: Bengal, 1890–1940*, Princeton: Princeton University Press, 1989, p. 149.
22. S. K. Gurtu, 'Hydro-electric Development', *ToI*, 4 Feb. 1926.
23. *Report and Proceedings of the Commission Appointed to Consider the Working of Factories in the Bombay Presidency, 1885*, Bombay: Government Central Press, 1885, pp. 57, 70–71. IOR, V/26/670/86.
24. *Report and Proceedings*, 1885, p. 47.
25. Lokhande's leadership, Sarkar argues, was more radical than that of earlier leaders, straddling two worlds—that of an emerging mass political activism and the older, familiar world of petitioning. See Aditya Sarkar, 'Regulated Labour, Unruly Workers: The Making of Industrial Relations in Late Nineteenth Century Bombay', Ph.D. thesis, SOAS, London, 2009, pp. 250–52.
26. By 'vitiated', he implied breathing in of the cotton dust.
27. *Report of the IFLC*, Vol. II, p. 61. In Ahmedabad, the collector reported that the dwelling houses of labouring classes were closed and confined, and if they compared the atmosphere of the mill with the atmosphere of these houses, there would not be very much difference. Most of the Ahmedabad mills were situated in open grounds and had plenty of space, and the more recent mills were all built on 'modern' lines. Dadabhai Nasarvanji Nanvati, city magistrate of Ahmedabad, however, argued that handloom weavers who worked from homes were healthier than mill workers.
28. Quoted in Sandip Hazareesingh, *The Colonial City and the Challenge of Modernity: Urban Hegemonies and Civic Contestations in Bombay City, 1900–1925*, Delhi: Orient Longman, 2007, p. 238.
29. 'Bombay's Slaves', *ToI*, 13 Sep. 1905. The language was particularly sharp regarding profit making, while ignoring other issues: 'Naturally the apologists for the worst features of the existing system are fertile in explanation and excuse. The evil accompaniment of money grubbing never lack defenders.' Later commentators have argued that this particular editorial hit hard, and mill owners and officials alike took keen note of it. See A. G. Clow, 'Indian Factory Legislation: A Historical Survey', *Bulletins of Indian Industries & Labour*, 37, Dec. 1926, p. 36.

30. Florence Barnes, 'Maternity Benefits to Industrial Workers: Final Report of the Lady Doctor', *Labour Gazette*, 2, 1, Sep. 1922; Dagmar F. Curjel, *Women's Labour in Bengal Industries*, Calcutta: Superintendent, Government Printing, 1923. Dr Barnes and Dr Curjel were officers of the Women's Medical Service whose services had been lent to the provincial governments by the Council of the Countess of Dufferin's Fund in 1921 to investigate the conditions of women's employment, with a special focus on the maternity question.
31. This remark was to be often cited, including by the Royal Commission on Labour in 1930, as a true description of conditions in Bombay's chawls, to make the argument that the working conditions in factories were more often considerably in advance of the conditions in the homes of the workers. Cited in Royal Commission on Labour in India, *Evidence*, Vol. I, Part 1, Bombay Presidency (Including Sind), London: His Majesty's Stationery Office, 1931, p. 27.
32. Curjel, *Women's Labour*, p. 14.
33. Curjel, *Women's Labour*, p. 15.
34. L. S. S. O'Malley, *City of Calcutta*, Census of India, 1911, Vol. VI, Part I, Report, p. 56.
35. Cited in O'Malley, *City of Calcutta*, p. 56.
36. The idea of 'dust everywhere' was commonplace enough, finding mention as much in travelogues as in public health literature. However, a distinction may be made between dust caused by the hoofs of animals or the sweeping of floors, on the one hand, and industrial dust, on the other. See Janet Kelman, *Labour in India: A Study of the Conditions of Indian Women in Modern Industry*, London: George Allen & Unwin, 1923.
37. In answering this question, she also wished to refute the argument sometimes offered that jute dust was not injurious as it fell rapidly. Instead, she pointed out, dust often fell on the faces of workers and on heads and bodies of the children who lay on the floors or in hammocks and that workers' own responses suggested that they considered it injurious. See Kelman, *Labour in India*, pp. 133–34. On the link between dust and tetanus among jute workers in Dundee, and enquiries regarding the same in India, see Harry Wilson, 'Jute', in Thomas Oliver, ed., *Dangerous Trades*, London: John Murray, 1902, pp. 650–62.
38. Kelman, *Labour in India*, pp. 133–34. This image of children being especially vulnerable was to persist, with observers noting that

the practice could not be condoned simply because dust was also present on streets and in bazaars, and equally injurious there. See Cecile Matheson, *Indian Industry: Yesterday, To-day and To-morrow*, London: OUP, 1930, pp. 52–53. The Royal Commission on Labour also observed that dust continued to be a health hazard in factories in India, being linked to pulmonary diseases. See Royal Commission on Labour in India, *Evidence*, p. 56.

39. Kelman, *Labour in India*, pp. 144–46. She does mention the presence of chimneys in some of the newer chawls, which might have helped in the dissipation of smoke from the rooms.
40. G. M. Broughton, *Labour in Indian Industries*, London: OUP, 1924, p. 108. Several others too commented on the continuing use of biomass—especially wood and dried cakes of cow dung—for heating and cooking purposes in lower-class Indian homes, with usually no provision made for the escape of the gases generated during this combustion, resulting in the air of such rooms becoming foul and poisonous. See C. L. Dunn and D. D. Pandya, *Indian Hygiene and Public Health*, Calcutta: Butterworth & Co., 1925, p. 47.
41. Ian Douglas, *Cities: An Environmental History*, London: I. B. Tauris, 2013, p. 71. This observation is with reference to nineteenth-century Manchester. In the USA too, such gendered impacts were common. See Angela Gugliotta, 'Class, Gender and Coal Smoke: Gender Ideology and Environmental Justice in Pittsburgh, 1868–1914', *Environmental History*, 5, 2, 2000, pp. 165–93
42. The estimate was based on an investigation of 20 homes of mill workers. A similar investigation was made of the homes of 70 women of the poorer non-industrial class. See Margaret Balfour and Shakuntla Talpade, 'The Maternity Conditions of Women Mill-Workers in India', *The Indian Medical Gazette*, May 1930, pp. 241–49.
43. A. R. Burnett-Hurst, *Labour and Housing in Bombay: A Study in the Economic Conditions of the Wage Earning Classes of Bombay*, London: P. S. King & Son, 1925, p. 19.
44. For a close reading of these assumptions in Burnett-Hurst's texts and images, see Juned Shaikh, 'Imaging Caste: Photography, the Housing Question and the Making of Sociology in Colonial Bombay, 1900–1939', *South Asia: Journal of South Asian Studies*, 37, 3, 2014, pp. 491–514.
45. Burnett-Hurst, *Labour and Housing*, p. 27.

46. Government of Bombay, *Report on the Working of the Development Directorate*, 1924. In the particular instance being described in this report, the Mahars were singled out for being insanitary.
47. G. R. Pradhan, *Untouchable Workers of Bombay City*, Bombay: Karnatak Publishing House, 1938, p. 12.
48. The lack of electricity in chawls and consequent harms to inmates on account of smoke nuisance also found some mention in the media. See 'Development Department Chawls', *BC*, 23 Jul. 1926.
49. Pradhan, *Untouchable Workers*, p. 22.
50. Pradhan, *Untouchable Workers*, p. 16.
51. A. C. Roy Choudhury, *Report on an Enquiry into the Standard of Living of Jute Mill Workers in Bengal*, Calcutta: Bengal Secretariat Book Depot, 1930, pp. 21, 25.
52. Roy Choudhury, *Standard of Living*, p. 21.
53. In a case of gas poisoning in Teli Mohalla in Bombay in 1933, a four-year-old child and her father were reported to have suddenly died while sleeping next to a *segree* that had been lit by the father and left as such through the night. The jury that enquired into the case returned the verdict that both the deaths were 'due to poisoning from gases generated from charcoal in a closed room'. See, 'Death from Gas Poisoning', *BC*, 5 Jan. 1933.
54. Calcutta Municipal Corporation, *Annual Report*, 1929–30, Vol. I, Part II, p. 6. Several diseases were grouped under this heading—acute bronchitis, chronic bronchitis, broncho-pneumonia and pneumonia.
55. BMC, *Annual Report*, 1926–27, Part II, p. 31; 1931–32, Part II, p. 41.
56. Calcutta Municipal Corporation, *Annual Report*, 1936–37, Vol. I, Part III, p. 28. Interestingly, in Bombay around the same time, we also begin to see links begin posited not only between smoke and health but more directly with the presence of coal itself. Thus, for instance, in June 1934, some residents of Grant Road and Dadar petitioned to the municipality that the unloading of charcoal in their locality was troublesome and injurious to their health. Municipal authorities in turn wrote to traders and to the Great Indian Peninsula (GIP) and Bombay, Baroda and Central India (BB&CI) railways that no unloading of coal would be allowed after the monsoon at any place other than at Haji Bunder at the Bombay Port Trust coal depot. See 'Coal: A Nuisance', *ToI*, 23 Jun. 1934.

57. Bengal SNC, *Annual Report*, 1926.
58. This does not imply, however, that domestic smoke was absent in Europe or the USA. Indeed, if anything, it remained a major problem right through the nineteenth and twentieth centuries in most homes on account of the burning of domestic fires for heat and comfort.
59. 'Indian Cooks and Cooking', *The Statesman*, 17 Jun. 1917. This report mentions the work of B. B. Bhattacharjee, district building surveyor, no. 11 district, in this regard.
60. Bengal SNC, *Annual Report*, 1928. Similarly, about Bombay, it was argued by some that tuberculosis could not be effectively cured in chawls as these were situated in areas exposed to chronic smoke nuisance, which, together with other factors, made medical improvement difficult. 'Problem of T. B. and How to Deal with It', *BC*, 20 Oct. 1940.
61. Frederick Grover, *Report on the Abatement of Smoke Nuisance in Calcutta*. NAI, Home, Public B, No. 95, Nov. 1904.
62. 'Smoky Bombay: A Cheap Preventive', *ToI*, 25 Feb. 1908.
63. Bengal SNC, *Annual Report*, 1906–07.
64. Bengal SNC, *Annual Report*, 1907–08. This in a city that was anyway lowly built, 'hugging the ground'. See Ira Klein, 'Urban Development and Public Health: Materialism, Environment and Mortality in Calcutta', in Giri Raj Gupta, ed., *Urban India*, Main Currents in Indian Sociology, Vol. VI, Durham: Carolina Academic Press, 1983, pp. 170–92. What such a comparison also implied was that smoke mattered only where people could breathe, not in the atmosphere as such. In Europe itself, this view would be challenged only much later when the phenomenon of acid rain came into prominence, suggesting the movement of polluted air from one airshed to another. See, Hugh Gorman, *Redefining Efficiency: Pollution Concerns, Regulatory Mechanisms and Technological Change in the US Petroleum Industry*, Akron, OH: University of Akron Press, 2001, pp. 316–17.
65. 'Calcutta Smoke Nuisance', as reproduced in 'Indian Notes and News', *ToI*, 13 Apr. 1909.
66. Bengal SNC, *Annual Report*, 1913–14.
67. Bengal SNC, *Annual Report*, 1917–18.
68. Bengal SNC, *Annual Report*, 1918–19.
69. Bengal SNC, *Annual Report*, 1920.

70. According to some newspaper reports, the extent of manufacture of coke from raw coal was also a function of price, with coke-making increasing with rise in coal prices in Bengal, though the commission itself did not mention price as a factor in the making of such choices. See 'Smoke Abatement', *ToI*, 14 Oct. 1921.
71. Bengal SNC, *Annual Report*, 1921.
72. Bengal SNC, *Annual Report*, 1923, 1925, 1926.
73. Bengal SNC, *Annual Report*, 1931.
74. Bengal SNC, *Annual Report*, 1928.
75. Bengal SNC, *Annual Report*, 1924.
76. Bengal SNC, *Annual Report*, 1927.
77. Uekötter suggests something similar for the USA—the more the emissions from trade and industry declined, the more obvious the domestic smoke problem became. Frank Uekötter, *The Age of Smoke: Environmental Policy in Germany and the United States, 1800–1970*, Pittsburgh: University of Pittsburgh Press, 2009, pp. 75–83. Nye makes the additional argument that in the USA, the focus shifted to smoke as a medical concern in the 1920s and 1930s once the efficiency approach lost credibility with reformers. David Nye, 'Smoke Gets in Your Eyes: Pollution, Aesthetics and Social Class', *Reviews in American History*, 28, 3, 2000, pp. 422–27.
78. The eventual remedy, they suggested, would be the use of smokeless gas appliances, which we consider later.
79. Bengal SNC, *Annual Report*, 1927.
80. Temperature inversions during the winter months, with incoming warmer air trapping cooler air below, was another reason for the worsening of the smoke pollution during this season, preventing the smoke from escaping. Low wind velocity compounded this problem regarding the dispersal of pollutants.
81. Bengal SNC, *Annual Report*, 1927.
82. Bengal SNC, *Annual Report*, 1929.
83. Bengal SNC, *Annual Report*, 1938. The link between tuberculosis and smoke, however, was not as direct as was being assumed. On the other hand, pneumonia, asthma and bronchitis could be directly linked to smoke. By the end of the 1920s, almost a third of the deaths among infants was being reported to be on account of bronchitis and broncho-pneumonia. One inference of this data could therefore be that smoke-induced ill health at worst affected the infant and the old, not so much people in the prime of their lives.

See Gugliotta, '"Hell with Lid Taken Off": A Cultural History of Air Pollution—Pittsburgh', Ph.D. diss., Notre Dame University, 2004. pp. 405–15.

84. Bengal SNC, *Annual Report*, 1927, 1928.
85. The main cause of domestic smoke in Calcutta, the commission observed in 1938, continued to be the use of cow dung cakes or wood scraps and kerosene oil for igniting soft coke which was almost universally used as fuel in household kitchens. Bengal SNC, *Annual Report*, 1938.
86. Bengal SNC, *Annual Report*, 1927. The average daily number of heaps of raw coal being made into coke was reported to have come down to six, as against about 600 in the days before action was taken. In another area near Telkul Ghat, owned by the East Indian Railway, it was estimated that about 4,000 tonnes of raw coal were annually made into coke. Bengal SNC, *Annual Report*, 1928.
87. Bengal SNC, *Annual Report*, 1930.
88. Bengal SNC, *Annual Report*, 1930, 1932.
89. Lt. Col. A. C. Chatterjee, 'The Smoke Demon of Calcutta', *ABP*, Health Supplement, 5 Feb. 1939. Chatterjee also drew attention to cow dung cakes being burnt near cowsheds for 'different purposes'. There were nearly 300 licenced cowsheds in the city, he pointed out, and possibly an equal number of unlicenced ones.
90. Apart from bodily harms, Chatterjee also drew attention to the tarnishing of furniture, blackening of the walls, fans and attire, and the corrosion of metals. Chatterjee, 'The Smoke Demon of Calcutta'.
91. The process was very simple, they wrote, involving strongly fanning the fire for three to four minutes when lighting—enough to assure that no smoke was emitted.
92. Bengal SNC, *Annual Report*, 1931.
93. The provision of a smoke flue for each hearth or group of hearths should also be insisted on, they wrote, before sanction was accorded to the plan of the house or bustee. Inspection of these houses after construction by responsible officers of the Health and Building Department of the corporation was also recommended, to ensure properly constructed chimneys before the bustees and houses were permitted to be occupied.
94. Bengal SNC, *Annual Report*, 1944.
95. Bengal SNC, *Annual Report*, 1938.

96. Bombay SNC, *Annual Report*, 1917–18, 1918–19.
97. Bombay SNC, *Annual Report*, 1917–18.
98. Bombay SNC, *Annual Report*, 1918–19.
99. Bombay SNC, *Annual Report*, 1918–19.
100. The concern in Bombay, as we saw in the previous chapter too, was almost obsessively about small industries and as regards ground smoke, having to do with locomotives and shipping. Surprisingly though, in Baroda, where there had been no great tradition of working towards abating smoke, a law was proposed in 1934 to control smoke nuisance in residential buildings by making it compulsory for all house owners to erect a specified kind of chimney or a window above the fireplace to prevent accumulation of smoke, designs for the same being supplied by local bodies. See 'Control of Smoke Nuisance: Proposed Legislation in Baroda', *ToI*, 28 Aug. 1934.
101. Bombay SNC, *Annual Report*, 1931–32. The association was also advised to encourage smoke nuisance complaints from the public and get them forwarded to the health officer or directly to the commission.
102. Gurtu, 'Hydro-electric Development'. Among the other advantages of electricity within homes, Gurtu mentioned cooking, lighting, electric fans, heating, ironing, grinding flour and knitting.
103. *BC*, 10 Feb. 1937.
104. *BC*, 6 Mar. 1937.
105. Publicity regarding the advantages of gas in domestic settings had started as early as 1854. See 'Advantages of the Use of Gas in Private Houses in Calcutta with a Description of the Manufacture of Coal Gas', Calcutta: Thacker, Spink & Co., 1854. Much of the advantage listed here had to do with lighting rather than smoke abatement. And even when publicised as beneficial from the perspective of lessening smoke, we may recall, coal gas could be considered clean only at the point of consumption, not at the source of production.
106. Bengal SNC, *Annual Report*, 1909–10.
107. Bengal SNC, *Annual Report*, 1910–1911, 1911–1912, 1912–1913.
108. Bengal SNC, *Annual Report*, 1917–18.
109. Bengal SNC, *Annual Report*, 1919–20.
110. Bengal SNC, *Annual Report*, 1920. In addition to the items mentioned the previous year, it also noted the provision of large cooking ranges.

111. Bengal SNC, *Annual Report*, 1921.
112. Bengal SNC, *Annual Report*, 1922.
113. The commission estimated that if these large buildings had instead used coal, there would have been a discharge of over half a tonne of soot. Bengal SNC, *Annual Report*, 1922.
114. Bengal SNC, *Annual Report*, 1923, 1924. Once again, the commission expressed satisfaction that not only did this prevent smoke that would have been emitted on account of the use of solid fuels, but it also prevented the discharge of over 5 tonnes of soot into the atmosphere.
115. Bengal SNC, *Annual Report*, 1925. However, it was also becoming obvious that many of these appliances were being fitted only into the new large buildings, and when introduced into smaller apartments, it could only be in more recent constructions.
116. Bengal SNC, *Annual Report*, 1925.
117. Bengal SNC, *Annual Report*, 1928.
118. Bengal SNC, *Annual Report*, 1930.
119. Reference was made to an article in the *Fuel Economist*, which referred to a project in Europe to collect the gas from coke ovens and distribute it to towns over 300 miles distant. The local coalfields of Calcutta, by comparison, were only 180 miles distant, where much gas was being wasted in the making of coke that could be profitably used for the needs of the city. Cited in Bengal SNC, *Annual Report*, 1927.
120. Bengal SNC, *Annual Report*, 1929.
121. Bengal SNC, *Annual Report*, 1931.
122. Chatterjee, 'The Smoke Demon of Calcutta'.
123. Bengal SNC, *Annual Report*, 1928.
124. Bengal SNC, *Annual Report*, 1927.
125. Bengal SNC, *Annual Report*, 1929.
126. Bengal SNC, *Annual Report*, 1929.
127. Bengal SNC, *Annual Report*, 1931.
128. Bengal SNC, *Annual Report*, 1930.
129. Globally, gas and electric companies seem to have competed for this market, with electric companies more successful in matters of lighting and power, while gas companies succeeded in cornering the market for cooking. See Graeme Gooday, *Domesticating Electricity: Technology, Uncertainty and Gender, 1880–1914*, London: Pickering & Chatto, 2008.

130. Government of Bengal, Commerce Department, *Report of the Committee Appointed to Enquire into the Charges for Electrical Energy Levied by the Calcutta Electricity Supply Corporation*, Alipore: Bengal Government Press, 1936. IOR, V/26/740/2.
131. As evidence of this, the enquiry mentioned that the company was able to pay a dividend of 3.5 per cent in 1900, the very first year of its functioning. See Government of Bengal, *Charges for Electrical Energy*, p. 78.
132. Government of Bengal, *Charges for Electrical Energy*, p. 80.
133. Government of Bengal, *Charges for Electrical Energy*, p. 46.
134. Government of Bengal, *Charges for Electrical Energy*, p. 56. This in a context where cooking by coal fires remained cheap.
135. Bengal SNC, *Annual Report*, 1932.
136. Bengal SNC, *Annual Report*, 1927, 1928.
137. Bengal SNC, *Annual Report*, 1928.
138. Bengal SNC, *Annual Report*, 1930. The section, they reported, was well patronised by the public. Such cooperation between anti-smoke organisations and gas companies had a long history in England, where beginning in the 1880s, gas companies had been giving considerable financial support to smoke abatement groups through sponsorship of exhibitions and conferences that would promote both their products and clean air. Such cooperation increased even further in the inter-war period when the industry increasingly used environmental claims to market itself. Peter Thorsheim, *Inventing Pollution: Coal, Smoke, and Culture in Britain since 1800*, Athens, USA: Ohio University Press, 2006, pp. 152–53.
139. Bengal SNC, *Annual Report*, 1932. Interestingly, the commission noted that two health exhibitions were also organised by them at Jamshedpur, at the request of Tata Iron and Steel Company Limited (TISCO), exhibiting charts and graphs regarding damages caused by smoke and possible remedies for the same. I would venture to add that the attitude to business to matters of environment, either aesthetic or grounded in efficiency or as matter of public health needs much greater elaboration. Globally, the first wave of writings on the subject had adopted a rather stark binary of evil industry versus people's health. Recent histories, however, present a more nuanced picture, highlighting the different interests at play across industries, their internal dynamics and the different motivations that prompted intervention. See Christine Rosen, 'Businessmen

against Pollution in Late Nineteenth Century Chicago', *Business History Review*, 69, 3, 1995, pp. 351–97.

140. Bengal SNC, *Annual Report*, 1933.
141. The publicity department had been created in 1932 and in addition to posters and handbills in Bengali, Hindi and Urdu, they also made use of magic lantern lectures, cinema shows and exhibitions.
142. Calcutta Municipal Corporation, *Annual Report*, 1938–39, Vol. I, p. 129.
143. 'Smoke Problem in Bengal', *ToI*, 13 Aug. 1936. By the end of the decade, the numbers seem to have come down a bit, less than half of the gas appliances installed in 1941 being for domestic purposes where they replaced older chulas. However, considering the limited numbers of new flats being built, the commission continued to suggest that the figure was satisfactory. Bengal SNC, *Annual Report*, 1941.
144. Bengal SNC, *Annual Report*, 1934.
145. Kabita Ray, *History of Public Health: Colonial Bengal, 1921–1947*, Calcutta: K. P. Bagchi & Co., 1998, p. 302.
146. Bengal SNC, *Annual Report*, 1934.
147. Suvobrata Sarkar, 'The Electrification of Colonial Calcutta: Role of the Innovators, Bureaucrats and Foreign Business Organization, 1880–1940', *Studies in History*, 34, 1, 2017, p. 70.
148. Bengal SNC, *Annual Report*, 1944.
149. Sarkar, 'Electrification of Colonial Calcutta', p. 75.
150. BMC, *Annual Report*, 1915–16, pp. 44–45.
151. A. T. Cooper, 'Recent Electrical Progress in India', *Journal of the Royal Society of Arts*, 77, 3994, 7 Jun. 1929, pp. 739–65.
152. Bombay SNC, *Annual Report*, 1932–33.
153. Bombay SNC, *Annual Report*, 1934–35.
154. Bombay SNC, *Annual Report*, 1931–32.
155. Bombay SNC, *Annual Report*, 1933–34.
156. BMC, *Annual Report*, 1942–43, p. 20.
157. For more on advertisements regarding domestic convenience, see Abigail MacGowan, 'Domestic Modern: Redecorating Homes in Bombay in the 1930s', *Journal of the Society of Architectural Historians*, 75, 4, Dec. 2016, pp. 424–46.
158. As is suggestive in these images, the introduction of gas came accompanied not only by the promise of greater cleanliness but also marked important transitions in the nature of domestic work, the making of the 'housewife' as a bourgeois person and a new

economy of the household. This merits a fuller study, but for similar experiences elsewhere in Asia, see Jordan Sand, *House and Home in Modern Japan: Architecture, Domestic Space and Bourgeois Culture, 1880–1930*, Cambridge, Massachusetts: Harvard University Press, 2003, especially Ch. 2, 'The Housewife's Laboratory'.

159. Indeed, even among middle-class households, such a transition to the use of gas and electricity for cooking and cleaning purposes would not happen till much later in the 1960s/1970s.

6 | Conclusion

Black smoke emitted into the atmosphere is matter in the wrong place, a waste of fuel, a public nuisance, and it ought to be prevented, or if produced consumed in the furnace ... There is no practical necessity for continuing the nuisance, and it ought not to be tolerated, because it is not only offensive, but it is doing daily, irreparable damage to public health, private and public property.[1]

The story of smoke in India is a fascinating study of the manner in which varied aspects of urban life, from production to consumption and from travel to habitation, together shape their environment. It is also an interesting matter with which to think simultaneously about two distinct spatial organisations—that of individual homes and factories, on the one hand, and that of the city and region, on the other, allowing us the possibility of enquiring into urban environments on different scales. It is also an interesting conceptual category—sometimes denoting waste, at other times, nuisance, its materiality as significant as its location. In this study, I have attempted to present an account of smoke that encapsulates these various dimensions, beginning with the different ways in which air was signified in colonial India, in sanitary discourses and in the discourse on air pollution, as relating to ordinary practices of cooking and dealing with refuse as also relating to modes of industrial production and fuel use. Energy, environment, resource conservation, spatial organisation and labour relations are thus drawn together, with smoke being simultaneously about waste (and therefore the proper utilisation of resources) and about pollution (necessitating disposal at a height). As such, it was also simultaneously about the city beautiful, about economic efficiency and about public health. And at all times, it was about reducing pollution by minimising waste, by making it profitable for industry to do so! To the extent that this was so, it was also less about moving industry, especially small-scale/traditional industry, out of the city, as enabling better environments through technological transformations and efficient fuel use within.[2] Beyond industry, dealing with smoke inside homes would also be less about their spatial location than about the use of alternative energy sources, with gas and electricity preferred over biomass and coal.

Industry, in and around Bombay and Calcutta, was both desirable, and problematic, in as much as growth and wealth came accompanied with filth and smoke. Much was said and written about filth, often attributing it to native habits, regarding which the colonial rulers had much to worry about. Filth was a marker of disease, which travelled easily from native habitats to cantonments, and not surprisingly, sanitation was an important axis of imperial governance. Within this story of sanitation, clean water took primacy, with purity of air being something of a secondary concern, acquiring significance from sometime around the 1880s. Sanitary streets, appropriate disposal of waste and slum clearance were some of the strategies adopted to address the unsanitary state of affairs, with free circulation of air and adequate sunlight as their goals. Alongside this, there was another discourse of sanitation, taking the factory, and not the city, as its site of intervention. This derived its significance from a consideration of conditions of work, especially of women and children, in which the moral energy to provide better conditions for Indian workers fused seamlessly with the trading interests of British manufacturers. Coupling the city and the factory together was an understanding of vitiated air, which could be improved through better ventilation, inside homes and factories, and outside on the streets. But while standards could be and were proposed for temperature and humidity inside factories, the air outside remained more descriptive, guiding policy interventions without quite mandating the desirable quality of air to be obtained. Similarly, while air inside homes could be vitiated, little to no attention was paid to how humidity and high temperature could also contribute to the burden of disease.

Architectural features such as the provision of windows and urban planning were of chief concern, and much faith was reposed in altered fuel use inside kitchens. Residence of the working classes, the chawls of Bombay and the bustees of Calcutta, were often commented upon in terms of the smoke that blighted their interiors, especially from the 1920s when links began to be forged between conditions inside homes and diseases suffered by workers. Factory owners were keen to establish that the mills as spaces of work were far healthier than the homes that had been disfigured by working-class Indians, more on account of their habits than for reasons of poverty, while labour activists and some among the community of researchers emphasised more the mismatch between wages and rentals that accounted for such abysmal living conditions. The general lack of sanitation having been already articulated, the focus would thus be on the Indian kitchen and the impossibility of smoke escaping from homes that had little direct opening to the outside.

Separate from these two accounts of the interiors of homes and factories was another account of smoke, polluted air that travelled widely over the city, blighting it in ways different from the filth encountered on streets. This was the story of smoke as 'waste'. Industrial smoke, while it signalled progress for some, also elicited an aesthetic dislike among many, especially the European administrators and elite Indians. This was Curzon's famous characterisation of Calcutta as a 'European city set down on Asiatic soil' and Sir Dorabji Tata's lament about the glories of Bombay if freed from smoke. The meteorological and the anthropogenic came together in this discourse, as determining the nature and effect of such smoke. Calcutta's temperature inversion was problematic, especially in the winters, as was the slow wind speed. Bombay, on the other hand, enjoyed the benefits of a sea breeze, though even this could not fully dissipate the smoke. Industries were differently located in and around the two cities, being more concentrated in the urban core in Bombay. And the sources of smoke differed too—while both would consider industrial and domestic smoke at some length, Calcutta seemed to have additionally laid emphasis on smoke from the making of coke in open air and from brick and lime kilns, while Bombay reported more regularly on smoke from locomotives and shipping, discharged closer to the ground. The kind of industries that demanded attention also seem to have differed from time to time, authorities in Calcutta emphasising smoke from the large industries, and occasionally from the oil mills and smaller establishments of north Calcutta, while Bombay seems to have focused far more on the problems of the smaller establishments located in dense residential neighbourhoods. And yet in both cases, much remained common—what was being addressed were visible signs of air pollution, viz., dense, black smoke, and it was addressed in manner that would not hamper industrial progress. The smoke nuisance could be tackled, because it was economically possible to tackle it, setting up a paradigm of regulation in which emphasis would be to 'protect the public', while simultaneously improving the position of the owners. Introducing a new Bill for smoke abatement in 1905, Government of India had thus justified its intervention by observing that 'the extra trouble and cost will be in great part repaid by the more economical working of the furnaces'.[3] In time, the Smoke Nuisances Commissions that came to be formed would equally emphasise on this cost-saving dimension, reiterating time and again their efforts at persuading industry that measures taken for smoke abatement reflected a saving to their firms rather than additional financial burden. Curzon himself was to reflect with some pride on the success of this strategy on his return to England:

> At first—and this is a difficulty that has to be met with here—the factory owners were, if not hostile, at any rate suspicious, but soon they came into line, and presently they gave us their hearty co-operation, no doubt influenced largely by the fact that we were able to show them that the expenditure in which they were involved was more than recouped by the saving in the fuel which they used.[4]

'Cost' itself, however, could be multiply conceived. Nicholson, for instance, had observed that excessive smoke was costly both to the manufacturer and the community, being a waste of fuel to the former and causing dirt and damage to the latter. Post war, the individual and the community got conjoined with the nation, as the wasteful use of coal was argued to represent the loss of a precious resource on which depended the future of the country. Given this context, good engineering and efficient use of resources attracted the maximum attention, with the imperative being to achieve only that which was feasible through the adoption of 'best practicable means'. So with aesthetic choices and economic reasoning came a faith in technical expertise and scientific design, which promised to offer the means by which smoke could be minimised, if not entirely eliminated. The use of electric power, drawn preferably from stations outside the city limits, in lieu of coal, was part of the same faith in industrial progress, prioritising clean consumption of energy over clean production, displacing the adverse impacts of smoke both in time and space by rendering it invisible.

This was the desire, common to Calcutta and Bombay, though the story played itself out rather differently in the two cities and with varied degrees of success. The initial responsibility for abating smoke in Calcutta was assigned to the Boiler Commission, headed by the commissioner of police, with magistrates having to work through the inspector appointed by them. Rather limited success was to be had in the first two to three decades, and the magistrate of Howrah often chaffed at the restrictions that he felt had been placed on his initiatives. But the commission remained convinced that not only did they have the monopoly of technically determining whether excessive smoke had been emitted or not but also that their mandate was to ensure that no injustice was done to industry. Industry, on the other hand, was far from convinced about the intention of the officials. The smoke abatement law of 1863, they argued, was 'illiberal', especially given that little to no technical support was being offered to them on how to abate smoke. By the concluding years of the nineteenth century, magistrates also seem to have changed their tune, increasingly unconvinced about the evidence offered by the smoke inspector, especially when counterpoised to the expertise of the mill manager. In this background, Lord Curzon, who had been appointed as the Viceroy in

1899 and who brought with him a new mission of administrative efficiency, decided to secure the assistance of an English expert to pass a new law and provide for a new administrative arrangement, one which would command greater authority among magistrates. The new law was passed in 1905, replacing the Boiler Commission with a Smoke Nuisances Commission and replacing the 'morally upright inspector with a keen eye' with a technically competent engineer, who would also be a person of virtue and so command the respect both of mill managers and of the magistrates. More importantly, it substituted the vague phrasing of the earlier law, which had emphasised due care being taken while lighting fires and use of best available technology with a clear limit on the maximum amount of smoke that could be emitted by each chimney, measured in minutes per hour, irrespective of the technology used.

In Bombay, things moved as slowly, and somewhat differently, with action against violators being initiated by the municipal commissioner, citing provisions of municipal law that were identical to the law of 1862 that prohibited excessive smoke in the city. However, as in Calcutta, prosecutions remained rather rare and the observations rather amateurish, and consequently often disputed in court settings. The extent of excessive smoke was determined, again as in Calcutta, by comparison—whether other industries on the same date had emitted less or whether the industry that was being prosecuted had, on other occasions, not smoked to the same extent, thereby suggesting carelessness on the day that excessive smoke was observed. Mill owners, on the other hand, drew attention to the use of inferior coal, when compared to mills in England, and also the fact that the municipality was as much a failure in controlling smoke from its own establishments, such as the mint, as it accused industry of. Like their counterparts in Calcutta, they too pleaded for greater technical support, pointing to the friendly relation of smoke inspectors and mill management in the USA. The health officer of the municipality, in turn, bemoaned the lack of specific time limits that allowed offending mills to escape lightly by emphasising the use of latest technology and having taken all possible precaution in the course of firing.

The *Times of India*, which seems to have run a sustained campaign on the issue, sought yet another kind of intervention, encouraging industry to adopt easily available technologies for smoke abatement. Thus, mill owners wanted the government to offer advice instead of prosecuting them; a section of the press urged them to make better investments and the municipality sought greater powers. However, as things developed, the municipality was to come out the loser, with a Smoke Nuisances Commission established in Bombay in 1912 instead, along lines similar to Calcutta, one in which officers of the municipality found no representation.

The laws having been amended, both in Calcutta and Bombay the commissions were quick to pronounce success, detailing the extent to which their advice had helped these cities to reduce smoke. There were some prosecutions too, especially in Calcutta, but in both cities, the commissions felt that they had achieved much more by partnering with industry rather than by prosecuting them. This, however, was a partial view at best, with responsibility for any excessive smoke that remained often passed on to the stokers who were increasingly subject to disciplinary action. Thus, while training and demonstration effects were emphasised for mill managers, who were without exception Europeans, firemen, who were exclusively poor Indians, were increasingly subject to fines and dismissals. This dual approach was to persist in the decades ahead too, with a brief interlude of attempted training of stokers, which proved an abysmal failure in both cities. Responses of the labour unions also seem to have been focused entirely on other issues, especially concerning hours of work and wages, to the exclusion of environmental concerns. In brief, workers were placed is dual settings—as residents, they could expect some concern about the amelioration of their living conditions. As labour, they could draw some empathy regarding 'conditions of work', including the state of the atmosphere inside factories. However, as stokers, a very specific kind of worker whose task was to provide the fire power that would generate steam and energy, they were often painted as careless and therefore deserving of punishment. Spaces of the interior and that of the city were thus framed differently, at the intersection of which were placed the workers, living in abysmal conditions, on the one hand, and subject to surveillance and discipline, on the other. By contrast, the commissions approached industry in an increasingly cooperative spirit, foregrounding an 'educative' approach towards owners and managers post First World War, almost entirely to the exclusion of punishment. In this, the commissions agreed not only to serve as advisors, as before, but also as providers of necessary expertise in terms of furnace and chimney design, as regards combustion engineering and low temperature carbonisation, imparting the latest knowledge on fuel use and modernising the administrative arrangements such that the task of 'mere observation' was passed on to amateur smoke observers, affording greater time to the engineer/smoke inspector to devote to technical matters. In Calcutta, beginning in the mid-1920s, the fact that over several consecutive years there had been no prosecution of mill owners was mentioned with great pride. In Bombay, prosecutions had been rather few and far between anyway. On the other hand, while seemingly far less invested in terms of collaborations with fuel research institutes abroad as was Calcutta, Bombay could take credit for conducting the first scientific study of the contribution of different sources to harmful particles in the air, initially in Bombay and then in Ahmedabad.

Success or otherwise, however, remained difficult to measure. The commissions took great pride in their achievements, and they made special mention of the extent to which their work had featured in the European press and among the European public. But a closer reading suggests that figures for minutes of smoke within an hour seemed to vacillate across years, especially in Bombay where the relative prices of different fuels—coal (imported from Wales or South Africa or brought from Calcutta), oil and hydroelectric power—seem to have had a greater impact than in Calcutta. Substitutions of one fuel source with another were reported rather frequently, but the use of oil by smaller establishments was seen as unsustainable, these being encouraged to use coke instead. In Calcutta, where the decline in smoke seems more steady, the problem that seems to have reappeared was of a different kind. After years of 'no prosecution', not only did the number of prosecutions rise in the 1930s, suggesting the limits of the 'advice alone' strategy, but also the evidence of the smoke observers, lacking technical expertise, again began to fail to find credibility with the magistrates, returning things to a position that was not very different from that at the end of the nineteenth century when the first amendments in law had been suggested.

Common to both Calcutta and Bombay, however, was another limitation that had to do with domestic smoke. When matters of the sort had been first brought to the notice of the commissions, they had merely acknowledged this source while admitting their inability to intervene, given that domestic smoke had been kept out of the purview of law. However, in time, the seriousness of the issue was recognised, especially from the 1920s, when the commissions felt that industrial smoke was more or less under control. Domestic smoke was troublesome for two reasons: (i) unlike industrial smoke, which could wane and wax on account of trade cycles, domestic smoke was constant, and (ii) unlike in Europe, where domestic smoke found an outlet through chimneys, in India, this tended to be released close to the ground, mixing there with other low-level smoke such as that emitted by locomotives. Solution in this regard, however, was more difficult to come by, given the number of households. An alternative therefore could only be had by the substitution of coal used inside homes with gas and electricity. Publicity about making such a switch thus came to assume greater primacy. We have already seen that domestic smoke, as a feature of working-class lives, had been noted as a problem inside chawls and bustees by those interested in labour issues. But domestic smoke, as an *urban* issue, was to be oriented differently, being addressed to owners of independent homes and newly constructed apartments buildings, where alone such a transition could be feasibly made. Private manufactures, aspiring for new markets, also addressed their communication to the same middle and upper classes. Just as the transition inside factories was to be made by enlightened owners and managers, with workers subject to fines and disciplining, so also in the case of domestic smoke, the promise of progress seemingly rested with the more affluent, who alone were seen as capable of making the new investments.[5] Workers could be considered vulnerable to smoke but rarely as the agents of making smoke-free cities. The middle class, on the other hand, by adopting gas and electricity, could hold on to the promise of becoming truly urban and serve as agents of the city beautiful and healthy!

This is where matters rested till the onset of the Second World War when, as in the case of the First World War, smoke abatement efforts slowed down in a bid to aid industries engaged in war-time production. The question of improvement of urban environment remained, to varying degrees of success, a technical exercise, with little prospect of democratic participation in either determining which issues mattered more or for what reason. The choice of addressing either large industries or small, considering homes more important than streets or otherwise, laying greater focus on training or on prosecution, all such choices lay entirely with the commissions. Indeed, even the press, which seems to have been active in highlighting the question of smoke till the first decade of the twentieth century, increasingly restricted itself to reporting the achievements of the commissions as mentioned in their annual reports. The public could petition the commission about issues that irked them, and indeed they did, more so in Calcutta than in Bombay, but they had to rest content with the assurance that the commission would look into their complaints and the declaration that the commission worked 'in the interest of all', by which they signalled smoke's burden as being generalised and universal. This claim, I suggest, must be read with some scepticism, remaining alert to the differential impacts of smoke across class and gender, among women and workers, inside chawls and in bustees and out on the streets, being linked to general poverty in some instances and with newer consumption and production patterns and energy use in others. Even more, smoke was not all that there was to air pollution; it was merely the most visible.[6]

The multiplicity of sources and sites, of practices ranging from cooperation and efficiency to discipline and punishment, of aesthetic choices and health burdens, make evident that to think of air today, as it was in the past, necessitates the fusing together of diverse issues, from resource conservation to pollution abatement and from conditions of work to democratic politics. Equally, it invites us to study air and its pollution in comparison, between mills in the nineteenth and twentieth centuries, and between cities as they compete for national and global rankings today. Such inter-city comparisons were not entirely missing in the past, with attention drawn to the similarities and differences not only between Calcutta and Bombay but also between them and London and Chicago. However, it would be no exaggeration to suggest that today such comparisons have acquired an entirely new significance, defining more than other elements, the attractiveness or otherwise of particular cities. Further, it would be no exaggeration to suggest that boosterism apart, we have also finally begun to acknowledge the extent to which our lives, and the life of the planet, depends on how current and past practices have shaped the atmosphere. And this concerns not only the most visible, viz., smoke, but also invisible particles and gases.[7] The intermittent, ad hoc attempts at redeeming the quality of air in the nineteenth and early twentieth centuries through publicity and persuasion have also been subject to diverse assessments, with several scholars concluding that this was the best possible strategy under the circumstances of the time.[8] On the other hand, given available scientific and technological knowledge, it is possible to argue that far more could have been done, with more stringent law keeping pace with technological advancements.[9] In the specific Indian context, and of colonial contexts in general, it is also possible to suggest that limited success apart, anti-smoke measures could display both racial and class biases, favouring Europeans as managers and inspectors, subjecting managers to occasional fines in instances of excessive smoke, while subjecting stokers to a more regular regime of fines and punishments.[10] There is also the matter of choice concerning the relative prioritisation of indoor and outdoor pollution, and the implications of such choices. Regional and global dimensions matter too. It was believed then, as it still is in some quarters still, that the way to deal with waste and pollution is to disperse it, to banish it from its immediate vicinity, to relocate industries or rearrange travel patterns. It did not quite work then, save to create other communities of harm. It does not quite work now, with more recent reports on air pollution on account of crop burning in the countryside or as it moves from more industrialised cities into neighbouring towns and the countryside, drawing attention to the *circulation* of air. National boundaries are no greater a barrier either, with winds from the Gulf and from neighbouring countries seen to carry dust and pollution into India, the extent of it changing with the seasons.[11] Air moves, travels and circulates, inside spaces and outside, in the city and in the suburbs. An engagement with it thus requires an engagement that is simultaneously place specific and fluid, attempting to reduce the extent of pollution at source as also to manage it across larger geographies. It requires that folded, inside spaces remain matters of concern—both the interiors of work spaces and homes—as also spaces that lie beyond administrative boundaries. To imagine otherwise, that is, to imagine an engagement with air that is confined only to the *public* arena of *bounded* metropolitan spaces, in a manner analogous to dealing with smoke in Bombay and Calcutta, Howrah and Ahmedabad in the nineteenth and twentieth centuries, would be to remain within the limits of twentieth-century urbanism.[12] Equally, to imagine air talk today as largely a dialogue of experts, of difference in strategy to be resolved through a rough consensus between the same experts, is to miss entirely the social nature of air, the manner in which issues of class, gender and spatial location are imbricated in the imaginations of clean and healthy cities. A 'clean' city, we suggest, demands an environmental ethics that combines efficiency of production with sensitivity to environmental standards and an appreciation of the social distribution of harms, to be arrived at through democratic deliberations. Then, and then only, would we have stepped beyond the colonial modernist and technocratic imaginary.

Notes

1. William Nicholson, *Report on Smoke Nuisances and their Abatement in Calcutta,* Calcutta: Bengal Secretariat Book Depot, 1906. IOR, V/27/780/14.

2. This duality of improving urban environments through relocating industry and/or by improving processes and refining technology stays with us till date, each with its own promise and limits.
3. *Proceedings of a Meeting of the Council of the Lt. Governor Held on 7th Jan 1905*. IOR, L/PJ/6/722 file 1381.
4. *Hansard*, 24 Mar. 1914.
5. To recollect, two distinct scenarios for energy transition developed within domestic settings, with gas and electricity being advocated largely for the use of the middle class and the rich and the use of coke instead of biomass suggested for the poor.
6. Stradling and Thorsheim draw our attention to the fact that in considering smoke, it was dark particulate emissions of fires that were most of concern, ignoring other gases such as sulphur dioxide. See David Stradling and Peter Thorsheim, 'The Smoke of Great Cities: British and American Efforts to Control Air Pollution, 1860–1914', *Environmental History*, 4, 1, 1999, pp. 6–31.
7. See Government of India, Ministry of Environment and Forests, *Policy Statement for Abatement of Pollution*, 1992, New Delhi.
8. Needless to add, all such assessments are with respect to developments in Europe and the USA. For a synoptic view of such positive assessments in England, see Stephen Mosley, *The Chimney of the World: A History of Smoke Pollution in Victorian and Edwardian Manchester*, Abindon, Oxon: Routledge, 2008, pp. 183–85. Mosley's own view is that far more could have been done than was achieved. Regarding the USA, Gorman too argues that the efficiency-oriented pollution control ethic could only achieve temporary success, before its eventual collapse by the last quarter of the twentieth century when there was a transition to a new ethic rooted in achieving socially determined environmental objectives. See Hugh Gorman, *Redefining Efficiency: Pollution Concerns, Regulatory Mechanisms and Technological Change in the US Petroleum Industry*, Akron, OH: University of Akron Press, 2001, p. 8.
9. This did not quite happen, and Mosley suggests that far from strengthening the legal regime, it was actually weakened, yielding poor results. On the other hand, Uekötter makes the case that 'few approaches have a worse record historically than litigation, and environmental historians are well advised to emphasize this point', going on to add that 'revealingly, the Progressive Era's antismoke movement did not make significant headway until it scaled down its aggressive rhetoric, turning from strict prosecution to education and

consultation with smoke generating businesses'. Frank Uekötter, *The Age of Smoke: Environmental Policy in Germany and the United States, 1800–1970*, Pittsburgh: University of Pittsburgh Press, 2009, pp. 11–12. In this context, it would be of interest to consider the regulation of urban environments through the Supreme Court of India and the National Green Tribunal in recent years. For some recent assessments of the same, see Geentanjoy Sahu, *Environmental Jurisprudence and the Supreme Court: Litigation, Interpretation and Implementation*, Delhi: Orient BlackSwan, 2014; Anuj Bhuwania, *Courting the People: Public Litigation in Post Emergency India*, Delhi: Cambridge University Press, 2017.

10. The working classes in England and the USA were not too enthused by anti-smoke campaigns either, fearing loss of jobs. In the Indian context, we do not quite know their views. As we suggest in earlier chapters, what we do have evidence for is greater willingness on the part of regulatory authorities to penalise workers for smoke and in general their prejudiced opinion about Indian labour.
11. 'Neighbouring States Responsible for 60 pc of Delhi Pollution: Study', *Indian Express*, 6 Oct. 2016; 'From Oman to NCR Towns, 74% of Delhi's Pollution Comes from Outside: Study', *Hindustan Times*, 16 Aug. 2018.
12. The suggestion, in other words, is in relation to both the specificity of a place and to the wider geographies in which a particular site is imbricated.

Bibliography

Archives

National Archives of India, Delhi

Home (Political)
Home (Sanitary)

West Bengal State Archives, Kolkata

Judicial Proceedings
Municipal Proceedings

Maharashtra State Archives, Mumbai

Development Department
General Department

India Office Records, UK

Online Resources

Internet Archive, https://archive.org.

Digital Repository of the Gokhale Institute of Politics and Economics, http://dspace.gipe.ac.in/xmlui/.

Sanjeevani Granth, Mumbai Asiatic Society, https://granthsanjeevani.com/jspui/.

Government Reports

Bengal, *Annual Report on the Administration of the Police for the Town and Suburbs of Calcutta for 1865–66.*

Bengal (India), *Review of the Industrial Position and Prospects in Bengal in 1908 with Special Reference to the Industrial Survey of 1890*, Part II, Calcutta: Bengal Secretariat Book Depot, 1908.

Bengal Retrenchment Committee Report, 1932.

Bengal Smoke Nuisances Commission, *Annual Report,* various years.

Bombay, *Provincial Report on the Working of the Indian Factories Act in the Bombay Presidency*, various years.

Bombay Government Gazette, 23 Feb. 1911, Part V.

Bombay Government Gazette, 31 Jan. 1912, Part VII.

Bombay Improvement Trust, *Annual Administration Report*, various years.

Bombay Municipal Corporation, *Annual Administration Report,* various years.

Bombay Smoke Nuisances Commission, *Annual Report,* various years.

Calcutta Improvement Trust, *Annual Administration Report*, various years.

Calcutta Municipal Corporation, *Annual Administration Report,* various years.

Electrical Undertakings in India (corrected up to 1916–17), Calcutta: Superintendent Government Printing, 1918. IOR, V/27/740/5.

Government of Bengal, Commerce Department, *Report of the Committee Appointed to Enquire into the Charges for Electrical Energy Levied by the Calcutta Electricity Supply Corporation*, Alipore: Bengal Government Press, 1936. IOR, V/26/740/2.

Government of Bengal, *Proceedings of the Council of the Lieut.-Governor of Bengal*, various years.

Government of Bengal, *Report of the Committee Appointed to Enquire into the Charges for Electrical Energy Levied by the Calcutta Electric Supply Corporation*, Alipore: Bengal Government Press, 1936. IOR, V/26/740/2.

Government of Bombay, *Bombay Legislative Council Debates*, various years.

Government of Bombay, *Proceedings of the Council of the Governor of Bombay*, various years.

Government of Bombay, *Provincial Report on the Working of the Indian Factories Act in the Bombay Presidency*, various years.

Government of Bombay, *Report on the Working of the Development Directorate for the Year Ending 31st March 1924*, Bombay: Government Central Press, 1924.

Government of India, *Electrical Undertakings in India* (corrected up to 1912), Calcutta: Superintendent Government Printing, 1913. IOR, V/27/740/4.

Government of India, *Indian Sanitary Policy, Being a Resolution Issued by the Governor General in Council, 23 May, 1914*, Calcutta: Superintendent Government Printing, 1914.

Government of India, Ministry of Environment and Forests, *Policy Statement for Abatement of Pollution*, 1992, New Delhi.

Government of India, *Report of Coalfields Committee*, Calcutta, Superintendent Government Printing, 1920.

Government of India, *Report of the Health Survey and Development Committee* (Bhore Committee), Delhi, 1946, two volumes.

Government of India, *Report of the Indian Industrial Commission 1916–18*, Calcutta: Superintendent Government Printing, 1918.

Grover, Frederick. *Report on the Abatement of Smoke Nuisance in Calcutta.* NAI, Home, Public B, No. 95, Nov. 1904.

Indian Industrial Commission, *Inspection Notes 1916–18*, Calcutta, 1918. IOR, Q/5/2.

Meares, J. W., *Triennial Report with a Preliminary Forecast of the Water Power Resources of India* (Hydroelectric Survey of India, Vol. III), Calcutta: Government Printing, 1922. IOR, V/27/740/11.

Mirams, A. E., *Plans and Specifications of Houses Suitable for Occupation by the Working Classes*, 2nd edition, Bombay: Government Central Press, 1925.

Nicholson, William, *Report on Smoke Nuisances and their Abatement in Calcutta*, Calcutta: Bengal Secretariat Book Depot, 1906. IOR, V/27/780/14.

Oldham, Thomas, *The Coal Resources and Production of India*, Calcutta: Office of Superintendent, Government Print, 1867.

O'Malley, L. S. S., *City of Calcutta*, Census of India, 1911, Vol. VI, Part I, Report, Calcutta: Bengal Secretariat Book Depot, 1913.

Ray, A. K., *A Short History of Calcutta Town and Suburbs*, Census of India, 1901, Vol. VII, Part I, 1902; reprint, Calcutta: Riddhi India, 1982.

Report and Proceedings of the Commission Appointed to Consider the Working of Factories in the Bombay Presidency, 1885, Bombay: Government Central Press, 1885. IOR, V/26/670/86.

Report of the Bombay Development Committee, Bombay: Government Central Press, 1914.

Report of the Bombay Retrenchment Committee, Poona: Yeravda Prison Press, 1923.

Report of the Commissioners Appointed by the Governor of Bombay to Enquire into the Condition of the Operatives in the Bombay Factories and the Necessity or Otherwise for the Passing of a Factory Act, 1875. Bombay: Government Central Press, 1875.

Report of the Committee on Industrial Unrest in Bengal, Calcutta: Bengal Secretariat Press, 1921.

Report of the Factory Ventilation Committee. Maharashtra State Archives, File No. 1232 of 1912, General Department, Vol. LXII.

Report of the Indian Factory Labour Commission, 1908, Simla: Central Branch Press, 1908.

Report of the Textile Factories Labour Committee, 1906, Bombay: Government Central Press, 1907.

Report of the Textile Labour Enquiry Committee, Vol. II, Final Report, Bombay: Government Central Press, 1940.

Roy Choudhury, A. C., *Report on an Enquiry into the Standard of Living of Jute Mill Workers in Bengal*, Calcutta: Bengal Secretariat Book Depot, 1930.

Royal Commission on Labour in India, *Evidence*. Vol. I, Part 1, Bombay Presidency (Including Sind), London: His Majesty's Stationery Office, 1931.

NON-GOVERNMENT REPORTS

Bombay Millowners' Association, *Annual Report*, various years.

BOOKS, DOCUMENTS AND ARTICLES

Advantages of the Use of Gas in Private Houses in Calcutta with a Description of the Manufacture of Coal Gas, Calcutta: Thacker, Spink & Co., 1854.

'Bernier's Travels in the Mogul Empire', *The Oriental Herald and Colonial Review*, Vol. X, 1826, p. 468.

Deaths in Bombay during 1863, Byculla, Bombay: Education Society's Press, 1864.

Allen, Michelle, *Cleansing the City: Sanitary Geographies in Victorian London*, Athens: Ohio University Press, 2008.

Anderson, M. R., 'The Conquest of Smoke: Legislation and Pollution in Colonial India', in David Arnold and Ramchandra Guha, eds, *Nature, Culture and Imperialism: Essays on Environmental History of South Asia*, Delhi: Oxford University Press, 1995, pp. 293–335.

Arnold, David, 'Burning Issues: Cremation and Incineration in Modern India', *NTM*, 24, 2016, pp. 393–419.

Arnold, David and Ramchandra Guha, eds, *Nature, Culture and Imperialism: Essays on Environmental History of South Asia*, Delhi: Oxford University Press, 1995.

Aspengren, Henrik, 'Empire: A Question of Hearts? The Social Turn in Colonial Government, Bombay, c. 1905–1925', in Mark Duffield and Vernon Hewitt, eds, *Empire, Development and Colonialism: The Past in the Present*, Woodbridge, Suffolk: James Curry, 2009, pp. 43–58.

Bagchi, Amiya, 'Wealth and Work in Calcutta, 1860–1921', in Sukanta Chaudhuri, ed., *Calcutta: The Living City*, Vol. I, Delhi: Oxford University Press, 1990.

Balachandran, Gopalan, 'Cultures of Protest in Transnational Contexts: Indian Seamen Abroad, 1886–1945', *Transforming Cultures eJournal*, 3, 2, 2008, pp. 45–75.

Balfour, Margaret and Shakuntla Talpade, 'The Maternity Conditions of Women Mill-workers in India', *The Indian Medical Gazette*, May 1930, pp. 241–49.

Banerjee, Sumanta, *The Parlour and the Streets: Elite and Popular Culture in Nineteenth Century Calcutta*, Calcutta: Seagull Books, 1989.

Barnes, Florence, 'Maternity Benefits to Industrial Workers. Final Report of the Lady Doctor', *Labour Gazette*, 2, 1, Sep. 1922.

Baviskar, Amita, 'Between Violence and Desire: Space, Power, and Identity in the Making of Metropolitan Delhi', *International Social Science Journal*, 55, 75, 2003, pp. 89–98.

Beattie, Martin, 'Colonial Space: Health and Modernity in Burra Bazar', *Traditional Dwellings and Settlements Review*, 14, 11, 2003, pp. 7–20.

Benjamin, N., 'Steam Boilers and Industrialization of Bombay, c.1850–1900: A Techno-economic History', *Proceedings of the Indian History Congress*, Vol. LXIX, 2008, pp. 612–24.

Bernier, François, *Travels in the Mogul Empire, 1656–1668*, translated and annotated by Archibald Constable, 1891, 2nd edition, revised by Vincent Smith, London: Oxford University Press, 1916.

Bhattacharya, Debjani, 'Interwar Housing Speculation and Rent Profiteering in Colonial Calcutta', *Comparative Study of South Asia, Africa and the Middle East*, 36, 3, 2016, pp. 465–82.

Bhuwania, Anuj, *Courting the People: Public Litigation in Post Emergency India*, Delhi: Cambridge University Press, 2017.

Bose, K. C., *Tuberculosis*, Appendix 21, Proceedings of the First All-India Sanitary Conference held at Bombay on 13 and 14 Nov. 1911, Calcutta: Superintendent, Government Printing, 1912, pp. 133–37.

Brimblecombe, P. and C. Pfister, eds, *The Silent Countdown: Essays in European Environmental History*, Berlin: Springer-Verlag, 1990.

Broughton, G. M., *Labour in Indian Industries*, London: Oxford University Press, 1924.

Brown, Michael, 'From Foetid Air to Filth: The Cultural Transformation of British Epidemiological Thought, c. 1780–1848', *Bulletin of the History of Medicine*, 82, 3, 2008, pp. 515–44.

Buchanan, D. H., *The Development of Capitalistic Enterprises in India*, New York: Macmillan, 1934.

Burke, Edmund, III, 'Preface', in Diana K. Davis and Edward Burke III, eds, *Environmental Imaginaries of the Middle East and North Africa*, Athens, USA: Ohio University Press, 2011.

Burnett-Hurst, A. R., *Labour and Housing in Bombay: A Study in the Economic Conditions of the Wage Earning Classes of Bombay*, London: P. S. King & Son, 1925.

Carter, Tim, 'Vitiated Air: A Victorian Villain?', *Journal of the Royal Society of Medicine*, 74, 12, 1981, pp. 914–19.

———, 'British Occupational Hygiene Practice 1720–1920', *Annals of Occupational Hygiene*, 48, 4, 2004, pp. 299–307.

Chakrabarty, Dipesh, *Rethinking Working Class History: Bengal, 1890–1940*, Princeton: Princeton University Press, 1989.

———, 'Of Garbage, Modernity and the Citizen's Gaze', *Economic and Political Weekly*, 7–14 Mar. 1992, pp. 541–46.

Chakraborti, Dipankar, 'Calcutta's Environment', in Sukanta Chaudhuri, ed., *Calcutta: The Living City*, Vol. II, Delhi: Oxford University Press, 1990, pp. 181–86.

Chandavarkar, Rajnarayan, *The Origins of Industrial Capitalism in India: Business Strategies and the Working Classes in Bombay, 1900–1940*, Cambridge: Cambridge University Press, 1994.

———, *History, Culture and the Indian City*, New York: Cambridge University Press, 2009.

Chang, Jiat-Hwee, '"Tropicalizing" Planning: Sanitation, Housing and Technologies of Improvement in Colonial Singapore, 1907–42', in Robert Peckham and David M. Pomfret, eds, *Imperial Contagions:*

Medicine, Hygiene and Cultures of Planning in Asia, Hong Kong: Hong Kong University Press, 2013, pp. 37–60.

Chatterjee, Partha, ed., 'Introduction', in *Texts of Power: Emerging Disciplines in Colonial Bengal,* Minneapolis: University of Minnesota Press, 1995.

Chaudhuri, K. N. and Clive J. Dewey, eds, *Economy and Society: Essays in Indian Economic and Social History*, Delhi: Oxford University Press, 1979.

Chaudhuri, Sukanta, ed., *Calcutta: The Living City*, Vols I and II, Delhi: Oxford University Press, 1990.

Choy, Timothy, *Ecologies of Comparison: An Ethnography of Endangerment in Hong Kong*, Durham: Duke University Press, 2011, pp. 121–52.

———, 'Air's Substantiations', in Kaushik Sunder Rajan, ed., *Lively Capital: Biotechnologies, Ethics and Governance in Global Markets*, Durham: Duke University Press, 2012.

Chu, Cecilia, 'Sanitation, Regulation and the Politics of Property in Colonial Hong Kong', in Robert Peckham and David M. Pomfret, eds, *Imperial Contagions: Medicine, Hygiene and Cultures of Planning in Asia*, Hong Kong: Hong Kong University Press, 2013, pp. 17–36.

Clow, A. G., 'Indian Factory Legislation: A Historical Survey', *Bulletins of Indian Industries & Labour*, 37, Dec. 1926.

Conybeare, Henry, *Report on the Sanitary State and Sanitary Requirements of Bombay*, Bombay: Bombay Education Society Press, 1855.

Cooper, A. T., 'Recent Electrical Progress in India', *Journal of the Royal Society of Arts*, 77, 3994, 7 Jun. 1929, pp. 739–65.

Cooper, Timothy, 'Modernity and the Politics of Waste in Britain', in Sverker Sorlin and P. Warde, eds, *Nature's End: History and the Environment*, London: Palgrave, 2009, pp. 247–72.

———, 'Recycling Modernity: Waste and Environmental History', *History Compass*, 8, 9, 2010, pp. 1114–25.

Corbin, Alain, *The Foul and the Fragrant: French Social Imagination*, New York: Berg Publishers, 1986.

Cotton, H. E. A., *Calcutta Old and New: A Historical and Descriptive Handbook to the City*, Calcutta: W. Newman & Co., 1907.

Curjel, Dagmar F., *Women's Labour in Bengal Industries*, Calcutta: Superintendent, Government Printing, 1923.

Datta, Partho, 'Patrick Geddes and the Metropolis', Nehru Memorial Museum and Library, Occasional Paper, History and Society, New Series, 5, 2012.

Datta, Partho, *Planning the City: Urbanization and Reform in Calcutta c. 1800–c.1940*, Delhi: Tulika Books, 2012.

Deaths in Bombay during 1863, Bombay: Education Society Press, 1864.

Dewey, Clive J., 'The Government of India's "New Industrial Policy", 1900–1925: Formation and Failure', in K. N. Chaudhuri and Clive J. Dewey, eds, *Economy and Society: Essays in Indian Economic and Social History*, Delhi: Oxford University Press, 1979, pp. 215–57.

Diana K. Davis and Edward Burke III, eds, *Environmental Imaginaries of the Middle East and North Africa*, Athens, USA: Ohio University Press, 2011.

Dossal, Mariam, *Imperial Designs and Indian Realities: The Planning of Bombay City, 1845–1875*, New York: Oxford University Press, 1991.

Douglas, Ian, *Cities: An Environmental History*, London: I. B. Tauris, 2013.

Duffield, Mark and Vernon Hewitt, eds, *Empire, Development and Colonialism: The Past in the Present*, Woodbridge, Suffolk: James Curry, 2009.

Dunn, C. L. and D. D. Pandya, *Indian Hygiene and Public Health*, Calcutta: Butterworth & Co., 1925.

DuPuis, E. Melanie, ed., *Smoke and Mirrors: The Politics and Culture of Air Pollution*, New York: New York University Press, 2004.

Eassie, William, *Cremation of the Dead: Its History and Bearings upon Public Health*, London: Smith, Elder & Co., 1875.

Edwardes, S. M., *The Rise of Bombay: A Retrospect*, Bombay: The Times of India Press, 1902.

Fischer-Tiné, Harald and Michael Mann, eds, *Colonialism as Civilising Mission: Cultural Ideology in British India*, London: Anthem Press, 2004.

Flanagan, Maureen, 'Environmental Justice in the City: A Theme for Urban Environmental History', *Environmental History*, 5, 2, 2000, pp. 159–64.

Fleming, James Rodger and Ann Johnson, eds, *Toxic Airs: Body, Place and Planet in Historical Perspective*, Pittsburgh: University of Pittsburgh Press, 2014.

Flick, Carols, 'The Movement for Smoke-abatement in 19th-century Britain', *Technology and Culture*, 21, 1, 1980, pp. 29–50.

Frasch, Tilman, 'Tracks in the City: Technology, Mobility and Society in Colonial Rangoon and Singapore', *Modern Asian Studies*, 42, 1, 2012: 97–118.

Gardner, Robert Dale, 'From Insurgency to Efficiency: The Smoke Abatement Campaign in Pittsburgh before World War I', *The Western Pennsylvania Historical Magazine*, 61, 3, 1978, pp. 187–202.

Ghosh, Anindita, *Claiming the City: Protests, Crime and Scandals in Colonial Calcutta, c.1860–1920*, Delhi: Oxford University Press, 2016.

Ghoshe, Babu Greesh Chunder, 'Female Occupations in Bengal', *Transactions of the Bengal Social Science Association*, Vol. II, Part I, Jan. 1868, pp. 125–34.

Gokhale, G. K., *Presidential Address to the Twenty First Session of the Indian National Congress*, Madras: Natesan and Company, 1905.

Gooday, Graeme, *Domesticating Electricity: Technology, Uncertainty and Gender, 1880–1914*, London: Pickering & Chatto, 2008.

Goode, S. W., *Municipal Calcutta: Its Institutions in Their Origin and Growth*, Edinburgh: T. and A. Constable, 1916.

Gopal, S., *British Policy in India, 1858–1905*, Cambridge: Cambridge University Press, 1965.

Gorman, Hugh, *Redefining Efficiency: Pollution Concerns, Regulatory Mechanisms and Technological Change in the US Petroleum Industry*, Akron, OH: University of Akron Press, 2001.

Gugliotta, Angela, 'Class, Gender and Coal Smoke: Gender Ideology and Environmental Justice in Pittsburgh, 1868–1914', *Environmental History*, 5, 2, 2000, pp. 165–93.

———, '"Hell with the Lid Taken Off": A Cultural History of Air Pollution—Pittsburgh', Ph.D. diss., Notre Dame University, 2004.

———, 'The "Smoky City" between the Wars', in E. Melanie DuPuis, ed., *Smoke and Mirrors: The Politics and Culture of Air* Pollution, New York: New York University Press, 2004, pp. 100–18.

Guha, Ramchandra and J. Martiner-Alier, *Varieties of Environmentalism: Essays North and South*, London: Earthscan, 1997.

Gupta, Giri Raj, ed., *Urban India*, Main Currents in Indian Sociology, Vol. VI, Durham: Carolina Academic Press, 1983.

Gupta, Nilanjana, ed., *Strangely Beloved: Writings on Calcutta*, Delhi: Rupa Publications, 2014.

Hamlin, Christopher, 'Surgeon Reginald Orton and the Pathology of Deadly Air: The Contest for Context in Environmental Health', in James Rodger Fleming and Ann Johnson, eds, *Toxic Airs: Body, Place and Planet in Historical Perspective*, Pittsburgh: University of Pittsburgh Press, 2014, pp. 23–49.

Harris, Richard, 'Development and Hybridity Made Concrete in the Colonies', *Environment and Planning A*, 40, 2008, pp. 15–36.

Harrison, Mark, *Public Health in British India: Anglo-Indian Preventive Medicine 1859–1914*, Cambridge: Cambridge University Press, 1994.

Hawes, Richard, 'The Municipal Regulation of Smoke Pollution in Liverpool, 1853–1866', *Environment and History*, 4, 1, 1998, pp. 75–90.

Hazareesingh, Sandip, *The Colonial City and the Challenge of Modernity: Urban Hegemonies and Civic Contestations in Bombay City, 1900–1925*, Delhi: Orient Longman, 2007.

Health Effects Institute, *The State of Global Air: Special Report*, 2019, Boston, MA: Health Effects Institute.

Hebbert, Michael, 'A City in Good Shape: Town Planning and Public Health', *Town Planning Review*, 70, 4, 1999, pp. 433–53.

Hewlett, T. G., *A Paper on the Sanitary State of Bombay*, 1869.

Heynen, Nik, Maria Kaika and Erik Sywngedouw, eds, *In the Nature of Cities: Urban Political Ecology and the Politics of Urban Metabolism*, London and New York: Routledge, 2006.

Home, Robert, *Of Planting and Planning: The Making of British Colonial Cities*, London: E & FN Spon, 1997.

Huang, Xuelei, 'Deodorizing China: Odour, Ordure, and Colonial (Dis)order in Shanghai, 1840s–1940s', *Modern Asian Studies*, 50, 3, 2016, pp. 1092–122.

Jackson, Lee, *Dirty, Old London: The Victorian Fight Against Filth*, New Haven and London: Yale University Press, 2014.

Joshi, Chitra, *Lost Worlds: Indian Labour and Its Forgotten Histories*, Delhi: Oxford University Press, 2003.

Kaika, Maria, *The City of Flows: Modernity, Nature and the City*, London & New York: Routledge, 2004.

Kelman, Janet, *Labour in India: A Study of the Conditions of Indian Women in Modern Industry*, London: George Allen & Unwin, 1923.

Kerr, Ian, 'Representation and Representations of the Railways of Colonial and Post-colonial South Asia', *Modern Asian Studies*, 37, 2, 2003, pp. 287–326.

Khan, Syed Sirdar Ali, *Lord Curzon's Administration in India: What He Promised, What He Performed*, Bombay: Times Press, 1905.

Kidambi, Prashant, *The Making of an Indian Metropolis: Colonial Governance and Public Culture in Bombay, 1890–1920*, Burlington, Vermont: Ashgate Publishing Company, 2007.

Kiechle, Melanie, *Smell Detectives: An Olfactory History of Nineteenth Century Urban America*, Seattle: University of Washington Press, 2017.

Kipling, Rudyard, 'A Real Live City', in *City of Dreadful Night*, New York: Alex Grosset & Co., 1899. First published in *Pioneer*, 2 March 1888.

Klein, Ira, 'Urban Development and Public Health: Materialism, Environ-ment and Mortality in Calcutta', in Giri Raj Gupta, ed., *Urban India*, Main Currents in Indian Sociology, Vol. VI, New Delhi: Vikas, 1983, pp. 170–92.

———, 'Urban Development and Death: Bombay City, 1870–1914', *Modern Asian Studies*, 20, 4, 1986, pp. 725–54.

Kling, Blair B., *Partner in Empire: Dwarkanath Tagore and the Age of Enterprise in Eastern India,* Berkeley: University of California Press, 1976.

Kreitman, Paul, 'Attacked by Excrement: The Political Ecology of Shit in Wartime and Postwar Tokyo', *Environmental History*, 23, 2, 2018, pp. 342–66.

Kumar, Deepak, ed., 'Colonial Science: A Look at the Indian Experience', in *Science and Empire: Essays in the Indian Context, 1700–1947*, Delhi: Anamika Prakashan, 1991, pp. 6–12.

———, ed., *Science and Empire: Essays in the Indian Context, 1700–1947*, Delhi: Anamika Prakashan, 1991.

Kumar, Radha, 'City Lives: Workers Housing and Rent in Bombay, 1911–1947', *Economic and Political Weekly*, 22, 30, 1987, pp. PE-47–PE-56.

Kydd, J. C., *A History of Factory Legislation in India*, Calcutta: University of Calcutta, 1920.

Lankester, Arthur, *Tuberculosis in India: Its Prevalence, Causation and Prevention*, Calcutta: Butterworth & Co., 1920.

Latham, Baldwin, *Report on the Sanitation of Bombay*, London: William Clowes & Sons Ltd, 1890.

Lipsett, H. Caldwell, *Lord Curzon in India, 1898–1903*, London: R. A. Everett & Co., 1903.

Luckin, Bill, '"The Heart and Home of Horror": The Great London Fogs of the Late Nineteenth Century', *Social History*, 28, 1, 2003, pp. 31–48.

MacGowan, Abigail, 'Domestic Modern: Redecorating Homes in Bombay in the 1930s', *Journal of the Society of Architectural Historians*, 75, 4, 2016, pp. 424–46.

Madureira, Nuno Luis, 'The Anxiety of Abundance: William Stanley Jevons and Coal Scarcity in the Nineteenth Century', *Environment and History*, 18, 3, 2012, pp. 395–421.

Mansell, James G., *The Age of Noise in Britain: Hearing Modernity*, Urbana: University of Illinois Press, 2017.

Martin, James Renald, *Notes on the Medical Topography of Calcutta*, Calcutta: Bengal Military Orphan Press, 1837.

Masselos, Jim, 'The Outside Inside: Incursions into the Marathi Household at the End of the Nineteenth Century', in *The City in Action: Bombay Struggles for Power*, Delhi: Oxford University Press, 2007.

Matheson, Cecile, *Indian Industry: Yesterday, To-day and To-morrow*, London: Oxford University Press, 1930.

Meade, Teresa and Mark Walker, eds, *Science, Medicine & Cultural Imperialism*, London: Macmillan, 1991.

Michael, L. W., *The History of the Municipal Corporation of the City of Bombay*, Bombay: Union Press, 1902.

Mieck, I., 'Reflections on a Typology of Historical Pollution: Complementary Conceptions', in P. Brimblecombe and Christian Pfister, eds, *The Silent Countdown: Essays in European Environmental History*, Berlin and Heidelberg: Springer-Verlag, pp. 73–80.

Minney, R. J., *Round About Calcutta*, London: Oxford University Press, 1922.

Misra, Maria, *Business, Race and Politics in British India, c. 1850–1960*, Oxford: Clarendon Press, 1999.

Mitchell, Timothy, ed., *Questions of Modernity*, Minneapolis: University of Minnesota Press, 2000.

Moore, Robin, 'Imperial India', in Andrew Porter, ed., *The Oxford History of the British Empire, Vol. III: The Nineteenth Century*, Oxford: Oxford University Press, 1999, pp. 422–46.

Moran, Michael, *The British Regulatory State: High Modernism and Hyper-innovation*, Oxford: Oxford University Press, 2003.

Moriss, Morris D., *The Emergence of an Indian Labour Force in India: A Study of the Bombay Cotton Mills*, Bombay: Oxford University Press, 1965.

Mosley, Stephen, *The Chimney of the World: A History of Smoke Pollution in Victorian and Edwardian Manchester*, Abindon, Oxon: Routledge, 2008, originally published by White Horse Press, 2001.

Nandy, Ashis, *An Ambiguous Journey to the City: The Village and Other Odd Ruins of the Self in the Indian Imagination*, Delhi: Oxford University Press, 2007.

Nye, David, 'Smoke Gets in Your Eyes: Pollution, Aesthetics and Social Class', *Reviews in American History*, 28, 3, 2000, pp. 422–27.

Orr, J. P., 'Light and Air in Dwellings in Bombay', lecture delivered before the Bombay Sanitary Association on 27 Jun. 1912, Bombay: Times Press, 1912.

Ottinger, Gwen, *Refining Expertise: How Responsible Engineers Subvert Environmental Justice Challenges*, New York: New York University Press, 2013.

Pande, Ishita, *Medicine, Race and Liberalism in British Bengal: Symptoms of Empire*, London: Routledge, 2010.

Peckham, Robert and David M. Pomfret, eds, *Imperial Contagions: Medicine, Hygiene and Cultures of Planning in Asia*, Hong Kong: Hong Kong University Press, 2013.

Platt, Harold, *Shock Cities: The Environmental Transformation and Reform of Manchester and Chicago*, Chicago: University of Chicago Press, 2005.

Polimeni, John M., Kozo Mayumi, Mario Giampietro and Blake Alcott, *The Jevons Paradox and the Myth of Resource Efficiency Improvements*, London: Earthscan, 2008.

Pradhan, G. R., *Untouchable Workers of Bombay City*, Bombay: Karnatak Publishing House, 1938.

Prakash, Gyan, 'Body Politic in Colonial India', in Timothy Mitchell, ed., *Questions of Modernity*, Minneapolis: University of Minnesota Press, 2000, pp. 189–222.

Rademacher, Anne, *Building Green: Environmental Architects and the Struggle for Sustainability in Mumbai*, Oakland, California: University of California Press, 2018.

Raleigh, T., *Lord Curzon in India*, London: Macmillan and Company, 1906.

Ramanna, Mridula, *Western Medicine and Public Health in Colonial Bombay, 1845–1895*, Delhi: Orient Longman, 2002.

Rao, Nikhil, *House, But No Garden: Apartment Living in Bombay's Suburbs, 1898–1964*, Minneapolis: University of Minnesota Press, 2013.

Ray, Animesh, 'The Calcutta Port', in Sukanta Chaudhuri, ed., *Calcutta: The Living City*, Vol. II, Delhi: Oxford University Press, pp. 123–27.

Ray, Indrajit, 'Dynamics of Bengal Coal Mining in the Nineteenth Century: Dissemination of Mineralogical Knowledge and Railway

Networking', *Indian Economic and Social History Review*, 52, 4, 2015, pp. 463–99.

Ray, Kabita, *History of Public Health: Colonial Bengal, 1921–1947*, Calcutta: K. P. Bagchi & Co., 1998.

Report of the First Industrial Conference Held at Poona, 1891, Poona: Arya Bhushan Press, 1891.

Rizvi, S. Ehtisham Husain, 'The Elements of Ghalib's Thoughts', trans. C. M. Naim and Muhammad Umar Memon, *Mahfil*, 5, 4, 1968/69, pp. 7–29, Asian Studies Center, Michigan State University.

Rogaski, Ruth, *Hygienic Modernity: Meanings of Health and Disease in Treaty-Port China*, Berkeley: University of California Press, 2004.

Rosen, Christine, 'Businessmen against Pollution in Late Nineteenth century Chicago', *Business History Review*, 69, 3, 1995, pp. 351–97.

Rosen, Christine and Joel Tarr, 'The Importance of an Urban Perspective in Environmental History', *Journal of Urban History*, 20, 3, May 1994, pp. 299–310.

Rosencranz, Armin and Michael Jackson, 'The Delhi Pollution Case: The Supreme Court of India and the Limits of Judicial Power', *Columbia Journal of Environmental Law*, 28, 2, 2003, pp. 223–54.

Rutnagur, S. M., ed., *Electricity in India: Being a History of the Tata Hydro-electric Project with Notes on the Mill Industry in Bombay and the Progress of Electric Drive in Indian Factories*, Bombay: Indian Textiles Journal, 1912.

———, *Bombay Industries: The Cotton Mills*, Bombay: The Indian Textile Journal Limited, 1927.

Sahu, Geetanjoy, *Environmental Jurisprudence and the Supreme Court: Litigation, Interpretation and Implementation*, Delhi: Orient BlackSwan, 2014.

Sand, Jordan, *House and Home in Modern Japan: Architecture, Domestic Space and Bourgeois Culture, 1880–1930*, Cambridge, Mass: Harvard University Press, 2003.

Sangwan, Satpal, 'Technology and Imperialism in the Indian Context: The Case of Steamboats, 1819–1839', in Teresa Meade and Mark Walker, eds, *Science, Medicine & Cultural Imperialism*, London: Macmillan, 1991, pp. 60–74.

Sarkar, Aditya, 'Regulated Labour, Unruly Workers: The Making of Industrial Relations in Late Nineteenth Century Bombay', Ph.D. thesis, SOAS, London, 2009.

Sarkar, Suvobrata, 'Domesticating Electric Power: Growth of Industry, Utilities and Research in Colonial India', *IESHR*, 52, 3, 2015: pp. 357–89.

———, 'The Electrification of Colonial Calcutta: Role of the Innovators, Bureaucrats and Foreign Business Organization, 1880–1940', *Studies in History*, 34, 1, 2017, pp. 48–76.

Searle, G. R., 'The Politics of National Efficiency and of War, 1900–1918', in Chris Wrigley, ed., *A Companion to Early Twentieth Century Britain*, Oxford, UK; Malden, MA: Blackwell Publishers, 2003, pp. 56–71.

Sen, Samita, *Women and Labour in Late Colonial India: The Bengal Jute Industry*, Cambridge: Cambridge University Press, 1999.

Shaikh, Juned, 'Imaging Caste: Photography, the Housing Question and the Making of Sociology in Colonial Bombay, 1900–1939', *South Asia: Journal of South Asian Studies*, 37, 3, 2014, pp. 491–514.

Sharan, Awadhendra, *In The City, Out of Place: Nuisance, Pollution, and Dwelling in Delhi, c.1860–c. 2000*, Delhi: Oxford University Press, 2014.

Short, Lisa Benton and John Rennie Short, *Cities and Nature*, Abingdon, Oxon: Routledge, 2008.

Simpson, W. J. R., *The Principles of Hygiene as Applied to Tropical and Sub-tropical Climates*, London: John Bale, Sons & Danielsson, 1908.

Sinha, Kaliprasanna, *The Observant Owl: Hootum's Vignettes of Nineteenth Century Calcutta* (*Hootum Pyanchar Naksha*, 1862), trans. Swarup Roy, Ranikhet: Permanent Black, 2008.

Sinha, Pradip, *Calcutta in Urban History*, Calcutta: Firma KLM Pvt. Ltd, 1978.

Sloterdijk, Peter, *Terror from the Air*, trans. Amy Patton and Steve Corcoran, Los Angeles, CA: Semiotext(e), 2009.

Sorlin, Sverker and P. Warde, eds, *Nature's End: History and the Environment*, London: Palgrave, 2009.

Srivastava, Priyanka, 'Creating a Healthy and "Decent" Industrial Labour Force: Health, Sanitation and Welfare in Colonial Bombay, 1896–1945', Ph.D. thesis, University of Cincinnati, 2012.

Sterndale, Reginald, *Municipal Work in India*, Calcutta: Thacker, Spink & Co., 1881.

Stradling, David, *Smokestacks and Progressives: Environmentalists, Engineers and Air Quality in America, 1881–1951*, Baltimore: Johns Hopkins University Press, 1999.

Stradling, David and Peter Thorsheim, 'The Smoke of Great Cities: British and American Efforts to Control Air Pollution, 1860–1914', *Environmental History*, 4, 1, 1999, pp. 6–31.

Sunder Rajan, Kaushik, ed., *Lively Capital. Biotechnologies, Ethics and Governance in Global Markets*, Durham: Duke University Press, 2012.

Tal, Alon, *Pollution in a Promised Land: An Environmental History of Israel*, Berkeley, CA: University of California Press, 2002.

Thorsheim, Peter, *Inventing Pollution: Coal, Smoke, and Culture in Britain since 1800*, Athens, USA: Ohio University Press, 2006.

Tyrell, Ian, *Crisis of the Wasteful Nation: Empire and Conservation in Theodore Roosevelt's America*, Chicago: University of Chicago Press, 2015.

Uekötter, Frank, 'Divergent Responses to Identical Problems: Businessmen and the Smoke Nuisance in Germany and the United States, 1880–1917', *Business History Review*, 73, 4, 1999, pp. 641–76.

———, 'The Merits of the Precautionary Principle: Controlling Automobile Exhausts in Germany and the United States before 1945', in E. Melanie DuPuis, ed., *Smoke and Mirrors: The Politics and Culture of Air Pollution*, New York: New York University Press, 2004, pp. 119–53.

———, *The Age of Smoke: Environmental Policy in Germany and the United States, 1800–1970*, Pittsburgh: University of Pittsburgh Press, 2009.

Upadhyay, Shashi Bhushan, *Existence, Identity and Mobilization: The Cotton Millworkers of Bombay, 1890–1919*, Delhi: Manohar, 2004.

Vernon, Rene, 'Remaking Urban Environments: The Political Ecology of Air Pollution in Delhi', *Environment and Planning A*, 38, 2006, pp. 2093–2109.

Vora, Rajendra, *The World's First Anti-dam Movement: The Mulshi Satyagraha 1920–1924*, Delhi: Permanent Black, 2009.

Wacha, D. E., *Shells from the Sands of Bombay*, Bombay: Bombay Chronicle Press, 1920.

Whitehead, Mark, *State, Science and the Skies: Governmentalities of the British Atmosphere*, Chichester, West Sussex: Wiley-Blackwell, 2009.

Wilhelm, Janine, *Environment and Pollution in Colonial India: Sewerage Technologies along the Sacred Ganges*, London: Routledge, 2016.

Wilson, Harry, 'Jute', in Thomas Oliver, ed., *Dangerous Trades*, London: John Murray, 1902, pp. 650–62.

Wrigley, Chris, ed., *A Companion to Early Twentieth Century Britain*, Oxford: Blackwell Publishers, 2003.

Zachariah, Benjamin, *Developing India: An Intellectual and Social History*, Delhi: Oxford University Press, 2005.

Zee, Jerry, 'Breathing in the City: Beijing and the Architecture of Air', *Scapegoat: Architecture/Landscape/Political Economy*, 8, 2015, pp. 46–57.

NEWSPAPERS

Amrita Bazar Patrika
The Bombay Chronicle
The Bombay Gazette
The Pioneer
The Statesman
The Times of India

Index